GRASS
the Forgiveness of Nature

GRASS

the Forgiveness of Nature

Charles Walters

Acres U.S.A.
Austin, Texas

Grass, the Forgiveness of Nature

The information in this book is true and complete to the best of our knowledge. Because the publisher cannot control field conditions and how this information is utilized, we disclaim any liability.

Acres U.S.A.
P.O. Box 91299
Austin, Texas 78709 U.S.A.
(512) 892-4400 • fax (512) 892-4448
info@acresusa.com • www.acresusa.com

Printed in the United States of America

Publisher's Cataloging-in-Publication

Walters, Charles, 1926-
Grass, the forgiveness of nature / Charles Walters. Austin, TX, ACRES U.S.A., 2006.

xxx, 306 pp., 23 cm., charts, tables.
Includes index.
ISBN: 0-911311-89-0

1. Pastures. 2. Forage plants. 3. Grazing.
I. Walters, Charles, 1926- II. Title.

SB199.L64 633.2'02

Dedicated to
grass-fed beef producers.

Contents

In Praise of Bluegrass

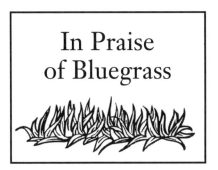

John James Ingalls was Senator from Kansas from 1873 to 1891. One of his speeches, printed in the Kansas Magazine *in 1872 and here reprinted because copies of it are hard to get, contains a passage that is quoted often. A fragment appears in* Grass: The Yearbook of Agriculture, 1948.

Attracted by the bland softness of an afternoon in my primeval winter in Kansas, I rode southward trough the dense forest that then covered the bluffs of the North Fork of Wildcat. The ground was sodden with the ooze of melting snow. The dripping trees were as motionless as granite. The last year's leaves, tenacious lingerers, loath to the leave scene of their brief bravery, adhered to the gray boughs like fragile bronze. There were no visible indications of life, but the broad, wintry landscape was flooded with that indescribable splendor that never was on sea or shore — a purple and silken softness, that half veiled, half disclosed the alien horizon, the vast curves of the remote river, the transient architecture of the clouds, and filled the responsive soul with a vague tumult of emotions, pensive and pathetic, in which regret and hope contended for the mastery. The dead and silent glove, with all its hidden kingdoms, seemed swimming like a bubble, suspended in an ethereal solution of amethyst and silver, compounded of the exhaling whiteness of the snow, the descending glory of the sky. A tropical atmosphere brooded upon an arctic scene, creating the strange spectacle of summer in winter, June in January, peculiar to Kansas, which unseen cannot be imagined, but once seen can never be forgotten. A sudden descent into the sheltered valley revealed an unexpected crescent of dazzling verdure, glittering like a meadow in early spring, unreal as an incantation, surprising as the sea to the soldiers of Xenophon as they stood upon the shore and shouted "Thalatta!" It was *bluegrass*, unknown in Eden, the final triumph of nature, reserved to compensate her favorite offspring in the new Paradise of Kansas for the loss of the old upon the banks of the Tigris and Euphrates.

Next in importance to the divine profusion of water, light, and air, those three great physical facts, which render existence possible, may be reckoned the universal beneficence of grass. Exaggerated by tropical heats and vapors to the gigantic cane congested with its saccharine secretion, or dwarfed by polar rigors to the fibrous hair of northern solitudes, embracing between these extremes the maize with its resolute pennons, the rice plant of southern swamps, the wheat, rye, barley, oats, and other cereals, no less than the humbler verdure of hillside, pasture, and prairie in the temperate zone, grass is the most widely distributed of all vegetable beings, and is at once the type of our life and the emblem of our mortality. Lying in the sunshine among the buttercups and dandelions of May, scarcely higher in intelligence than the minute tenants of that mimic wilderness, our earliest recollections are of grass; and when the fitful fever is ended, and the foolish wrangle of the market and forum is closed, grass heals over the scar which our descent into the bosom of the earth has made, and the carpet of the infant becomes the blanket of the dead.

As he reflected upon the brevity of human life, grass has been favorite symbol of the moralist, the chosen theme of the philosopher. "All flesh is grass," said the prophet; "My days are as the grass," sighed the troubled patriarch; and the pensive Nebuchadnezzer, in his penitential mood, exceeded even these, and, as the sacred historian informs us, did eat grass like an ox.

Grass is the forgiveness of nature — her constant benediction. Fields trampled with battle, saturated with blood, torn with the ruts of cannon, grow green again with grass, and carnage is forgotten. Streets abandoned by traffic become grass-grown like rural lanes, and are obliterated. Forests decay, harvests perish, flowers vanish, but grass is immortal. Beleaguered by the sullen host of winter, it withdraws into the impregnable fortress of its subterranean vitality, and emerges upon the first solicitation of spring. Sown by the winds by wandering birds, propagated by the subtle horticulture of the elements, which are its ministers and servants, it softens the rude outline of the world. Its tenacious fibres hold the earth in its place, and prevent is soluble components from washing into wasting sea. It invades the solitude of deserts, climbs the inaccessible slopes and forbidding pinnacles of mountains, modifies, climates, and determines the history, character, and destiny of nations. Unobtrusive and patient, it has immortal vigor and aggression. Banished from the thoroughfare and the field, it bides its time to return, and when vigilance is relaxed, or the dynasty has perished, it silently resumes the throne from which it has been expelled, but which it never abdicates. It bears no blazonry or bloom to charm the senses with fragrance or splendor, but its homely hue is more enchanting than the lily or the rose. It yields no fruit in earth or air, and yet should its harvest fail for a single year, famine would depopulate the world.

One grass differs from another grass in glory. One is vulgar and another patrician. There are grades in its vegetable nobility. Some varieties are useful. Some are beautiful. Others combine utility and ornament. The sour, reedy herbage of swamps is baseborn. Timothy is a valuable servant. Redtop and cover are a degree higher in the social scale. But the king of them all, with genuine blood royal, is *bluegrass.* Why it is called blue, save that it is more vividly and intensely green is inexplicable, but had its unknown priest baptized it with all the hues of the prism, he would not have changed its hereditary title to imperial superiority over all its humbler kin.

Tame, in his incomparable *History of English Literature*, has well said that the body of man in every country is deeply rooted in the soil of nature. He might properly have declared that men were wholly rooted in the soil, and that the character of nations, like that of forests, tubers, and grains, is entirely determined by the climate and soil in which they germinate. Dogmas grow like potatoes. Creeds and carrots, catechisms and cabbages, tenets and turnips, religions and rutabagas, governments and grasses, all depend upon the dew point and the thermal range. Give the philosopher a handful of soil, the mean annual temperature and rainfall, and his analysis would enable him to predict with absolute certainty the characteristics of the nation.

Calvinism transplanted to the plains of the Ganges would perish of inanition. Webster is as much an indigenous product of New England as its granite and its pines. Napoleon was possible only in France; Cromwell in England; Christ, and the splendid invention of immortality, alone in Palestine. Moral causes and qualities exert influences far beyond their nativity, and ideas are transplanted and exported to meet the temporary requirements of the tastes or necessities of man; as we see exotic palms in the conservatories of Chatsworth, russet apples at Surinam, and oranges in Atchison. But there is no growth: nothing but change of location. The phenomena of politics exhibit the operations of the same law.

The direct agency upon which all these conditions depend, and through which these forces operate, is food. Temperature, humidity, soil, sunlight, electricity, vital force, express themselves primarily in vegetable existence that furnishes the basis of that animal life which yields sustenance to the human race. What a man, a community, a nation can do, think, suffer, imagine or achieve depends upon what it eats.

The primary form of food is grass. Grass feeds the ox: the ox nourishes man: man dies and goes to grass again; and so the tide of life with everlasting repetition, in continuous circles, moves endlessly on and upward, and in more senses than one, all flesh is grass. But all flesh is not bluegrass. It if were, the devil's occupation would be gone.

Foreword

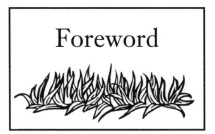

Edward O. Wilson, the Harvard expert on ants, also produced the world's best-known book on biodiversity. In it he implied that we need not focus on saving the planet. The planet will take care of itself if plants are allowed their service of providing oxygen to a threatened air supply. Pasture grass is dominant in sequestering carbon. It has been calculated by the Kika de la Garza Research Station of Westlaco, Texas, that increasing one-tenth of one percent humus to the soil in the fullness of time will sequester carbon not threatening us with global warming. It is the human race that needs saving. Therein lies the dilemma; to insure a future for mankind, more consideration has to be given to Earth's feelings, its heartbeat, its digestion. For less than even one military adventure, the rain forests of planet Earth could be set aside to salvage the planet's oxygen supply.

When I was a child, age 5 or 6, an uncle explained to me what fools "we mortals be," plowing up the native grasses of western Kansas. The words were hardly out of his mouth before the earth fielded an uptake, a magnified cloud of dust that followed an Erskine or Ford down a dusty road. That soil arose 10,000 feet in altitude; grounded airplanes, even grounded Charles A. Lindbergh, the newspapers said, as he crossed the Oklahoma Panhandle and southwestern Kansas.

These storms haunted more than they condemned. Life went on.

I was returned to the dust bowl and the issue of grass when Fletcher Sims of Canyon, Texas published a shocking report in the *Amarillo (Texas) Globe* entitled, "Goodbye, Ogallala, Hello Sahara Desert." He was writing about a water system that sat like a huge bath tub under a former native pasture area that ran north and south from South Dakota to Texas, east and west from a slice of Wyoming, Colorado and New Mexico into Kansas, Nebraska, South Dakota, Oklahoma and the Texas Panhandle. A tap into this reservoir did much to close down the dust cycle, but it was a short-sighted remedy.

Sims was writing about a landscape that was once an ecological climax call short-grass prairie. That grass once serviced buffalo herds that General Phil Sheridan defined as being several miles across. It sustained a cattle industry that required negligible quantities of water contained in the Ogallala and Santa Rosa Aquifers.

Wes Jackson of the Land Institute, Salina, Kansas, tells us that it took 500 years for the grasses above those aquifers to reach climax perfection. It took hardly a generation for the folly of plowmen to turn that soil upside down and produce abundant wheat crops.

Soon enough the agricultural economy forgot the disciplines that made crop production possible. At first farmers chose crops that could abide the discipline of sporadic rains. As the humus supply become eroded, the tap into the Ogallala, irrigation, plus the subsidy programs — even developers, Sims said — in effect put so many taps into the pristine water supply they promised near total exhaustion of the Ogallala in 18 years or less.

The climax crop that kept the soil in place was not simply grass. More correctly, it was forage composed of a dozen, even three or four-dozen species.

The high plains still had a fair measure of forage at the end of World War II. As more acres were plowed under for grain production, large feedlots came to Nebraska, Kansas, Colorado and Texas. Not only was the water being withdrawn from the Ogallala, now manure piled up. One of my first trips to the Texas Panhandle to see manure composting first hand allowed me to photograph a mountain of manure 40 to 60 feet high covering several acres of ground. One experiment buried 200 tons of manure per acre at coffin depth, totally ruining these acres for as long as it takes for the next ice age to arrive. Except for Sims, hardly anyone thought in terms of restoring organic matter to the soil.

As recently as the 1930s only a fraction of the high plains acres had been turned under, even though this insult activity brought on nature's revenge, Anhydrous ammonia had not made it to the high plains scene. Around 1939, the first of the great wells went down. Section after section still used to graze cattle became a miner's treasure house, extending the acres of crops thus produced. Soon enough half of the high plains became cultivated. As crop followed crop, yields started to decline. This defect was answered with more water and chemicals. As a consequence, the humus was hurried away repeatedly. "This reminds me of the horse trader trying to sell a cow to a farmer," said Fletcher Sims. "He was bragging on how much milk the cow produced. This led the farmer to question how much she ate. The trader explained that the best thing about her was that she sucked herself."

The more the soil was plowed and fertilized, the more inputs were required, "the problem feeds on itself," Sims pointed out.

There was a time when very little rainfall left the grass covered high plains. Now, because the soil has been wasted away, too much water separates. History buffs know about Charles Goodnight, how he put windmills and water tanks to work, producing free ranging cattle herds. Such ranching could have been sustained indefinitely without running a tap into the subterranean Ogallala. That Ogallala reservoir is seated some 200 feet below the surface.

Thus the questions of the hour, whether towns and farms depend on deep well irrigation are doomed as the Ogallala is sucked dry, or whether unsequestered carbon dioxide will ultimately make the planet a hothouse next to intolerable for flora and fauna — humans included. Standing between us and the pending legacy for our children is an appreciation of grass for areas that should be kept in grass, in a word, the pasture mystique.

Grass sequesters carbon and builds humus. Ranchers who keep good records tell us that herd grazing on fragile soils builds humus at the rate of a tenth of a percent per year, the exact amount necessary on all farmland to return a proper carbon dioxide blend.

In this survey of grass forage, if you wil, I have run a tap into the array of information contained in *Acres U.S.A.*, at conferences, in personal interviews, and in searches of pertinent and available literature. This is not an academic exercise. It is more a chat, the kind that takes place when eco-growers meet to cooperate with fellow growers and ranchers to help each other.

Here the case will be made for grass-fed beef, with side notes on the role of pasture in swine production and even range-fed chickens. Always, the emphasis is on low cost inputs, on repairing abused land, and on returning meat protein to a classic food status as opposed to the commodity fare it has become.

The primary source for each chapter has been clearly indicated, with four interviews preserved exactly as first recorded, questions and clear answers included. I have retained the most profound chapter of all until last for two reasons. It promises the most intelligent answer to healing the earth, and it is the most likely to be ignored by most Americans because it is somewhat esoteric and requires a break with pseudo-science and invites hard work.

Readers may wonder why some few pages of this volume have been devoted to brucellosis, Aftosa virus, bovine spongiform encephalopathy, and para-T, all syndromes that are cancelled out when pastures are flush with nutrients and animals have available their 25 cubic feet of air per minute, a blessing never available in confinement. The reader can come to an appropriate conclusion.

After the pages of this book were assembled, I e-mailed Hugh Lovel, formerly of Georgia, now farming in Queensland, Australia, for "a few comments." His response took my breath away, for which reason it is included in this foreword verbatim. It will surely put into focus the chapters that follow:

> Living in Australia, I see once again that two aspects of pasture management defeat graziers time and again in Australia as well as elsewhere, and mattes are only made worse by Australia's extended dry seasons and droughts. One is inappropriate grazing cycles, and the other is chemical fertility management.
>
> Pastures are best grazed hard but briefly in cycles of a month or more apart. This not only breaks up parasite reproduction but it allows for dense, robust regrowth.
>
> Grazing must be managed with an eye to maximizing forage growth (catching as much carbon as possible) as well as coverage of the soil surface (energy retention). As

Allan Savory aptly notes, aeration, organic matter and drainage all somewhat depend on soil cover.

Maximizing growth and soil cover go hand in hand, and the rule of thumb is to observe what works best under local conditions. Particularly in Australia conditions tend to be highly viable, so keen observation and flexibility are essential.

Too little grazing and vegetation grows up, blooms, makes seeds and just stands there in the way of catching more carbon. Growth stagnates and plant distribution thins out as regrowth is shaded by the dry, standing vegetation. I see this a lot where extended drought has reduced stocking rates. Then when good moisture is present pastures grow luxuriantly, top out, turn brown and stagnate. At that point, if stock were put on these pastures all they would have to eat is dry stems and seeds. This is a crisis that goes unnoticed, as there appears to be plenty of pasture available.

With too much grazing regrowth is nipped in the bud, plants run out of energy and soil cover becomes too sparse as carbon is not replenished as fast as it is digested. In both cases, the usual result is more, rather than less, bare soil. Energy is lost from bare soil, while it is drawn in by covered soil.

Continuous grazing of enclosed pastures tends to perpetuate parasites and diseases. Moreover, constant impact by heavy animals closes up the pores in the soil, reducing air and water intake. Even worse, when pasture plants are kept grazed down they never enjoy their growth potential. It is this growth potential, the ability plants have to catch carbon and turn it into organic compounds that can increase the amount of life on earth. Of course, this potential is improved by diverse, balanced mixtures of deep and shallow rooted grasses, legumes and forbs. (For those unfamiliar with the term, "forbs" means leafy, non-grassy plants such as chicories, dandelion, yarrows, comfrey, plantains, etc.)

The single most important aspect of managing vegetation is building up carbon. Carbon is the basis of life. Carbon is free. Sandy soils, rocky soils, dense clays and soils that are way out of mineral balance—all can be made productive as long as they contain high levels of organic, living carbon. Catching carbon and building organic matter is key to long-term success. No organic matter, no soil biology, no success.

It goes deeper than this, however. Carbon has strong affinity for hydrogen, which means carbon is a magnet for water. In photosynthesis, plants absorb light and combine carbon dioxide with water to build up sugars into complex organic compounds while releasing oxygen. This affinity of carbon for hydrogen affects rainfall.

Watch for the signs of this. Not only do forests and dense pastures cool the air and attract more rainfall, so do microorganisms in the soil. Carbon based material on the soil surface reduces the albedo, or reflective index. Litter left behind by brief but intensive grazing is a big improvement over bare soil. If the carbon on the surface is alive, absorbing energy and photosynthesizing, even it if is only algae, the cooling effect is even greater. Moisture-laden air masses passing over surfaces with a low albedo are much more likely to drop their moisture as rain.

It is important that there is some litter, however. Livestock are not the only grazers in the pasture. Litter provides habitat and fodder for myriad soil critters, of which earthworms, ants and dung beetles are some of the more noticeable types. These animals not only perforate the surface, burrow and aerate the soil, providing for better insoak of rain, they establish and maintain the abundant microbial growth that fixes nitrogen and provides minerals for pasture plants. Sometimes this means it is best to mow down pastures that have topped out and gone stagnant so the resulting litter can be used as fodder by the little soil-building critters who are not normally thought of as stock. Then regrowth is encouraged. This is fertilization without buying anything except tractor fuel and machinery maintenance.

What of these little grazers? Many of them stoke the soil biology, making it robust. Soil biology is the key to fertility. Earthworms and ants are two of the prominent players. Where earthworms primarily spread and nurture soil bacteria, ants are the premier fungal manipulators, and by and large it is fungi that access and release mineral nutrients such as calcium, magnesium, phosphorus and potassium.

Aside from putting poison on food, there has hardly been any worse blunder made in agriculture than thinking that one wants soil nutrients to be soluble. Even when well intentioned and unwitting, this doctrine is a deplorable lie, perpetuated by industries that may be sophisticated, but which have no more scruples than a drug pusher on the street corner. Unless you want nutrients to wash out when it rains, or collect in salt deposits where there is insufficient rain, you do not want nutrients to be soluble. What anyone can see if they only think about it is that nutrients should be insoluble but available. Insoluble-but-available nutrients are buoyed up against rains or droughts in the ever-conserving biology of the soil. Soil biology conserves fertility, and any "fertilizer" that kills off soil biology does not deserve to be called a fertilizer. This means anhydrous ammonia, urea, ammonium nitrate, muriate of potash (potassium chloride), superphosphate, monoammonium and diammonium phosphates and the like should be classified as soil killers rather than fertilizers, since they destroy the biology of the soil.

This story is never told by the chemical industries making these products, nor is it found in the media or the textbooks of universities dependent on industry for their funding. Harsh chemical fertilizers kill off soil biology, making growers dependent on a downward spiral of chemical fixes.

In the face of all the propaganda to sell chemicals as cheap, safe and effective, people have to do their own thinking before they realize that to achieve self-sufficiency they must improve soil biology. With the exception of independent consultants and those selling biological fertilizers, composts, compost tea, microbial innoculla, biodynamic preparations and the like no one tells this story. It's not hard to understand why because there is little or no money in it. If the fertilizer industry promoted good soil biology they would put an end to a business worth trillions of dollars worldwide.

Any chemical company agronomist who comes out, takes soil samples, analyzes them and recommends what fertilizers to apply at what rates, is there to sell his own products. It is understandable that every year he will hope to increase his sales. Using his services may seem ever so simple and carefree. One doesn't have to think twice. But is this sort of dependency the direction to go in? Slick corporations devoted to short-term profit deserve as much if not far more skepticism as snake oil salesman. Think about it.

Of course, graziers may simply accept their existing fertility levels and improve things gradually through grazing and recovery cycles, keeping good soil cover and maintaining a mix of deep and shallow rooted forage plans. This can be a successful, low input strategy. But judicious fertility improvements can shorten the path to robust growth from decades or centuries to a mere few years. There can be enormous payoffs from learning how to remineralize soils so that carbon, hydrogen, nitrogen, oxygen and sulfur are caught for free and retained.

For example, how long does it take for the biology of the soil to accrue sufficient molybdenum in the topsoil for robust nitrogen fixation? Decades? Centuries? Millennia? Molybdenum is the key nutrient for the enzyme, nitrogenase, which splits apart the triple bonded nitrogen gas from its inert atmospheric state. You won't get free nitrogen without nitrogenase, which is one of the first things to go down the tubes when nitrogen fertilizers are used. Only a couple ounces of molybdenum per acre, or 100-10 grams per hectare, are needed. I found this out, much to my surprise, my first year of farming. Whereupon I used molybdenum seed treatment powder. The cost was tiny, but the effect was huge. Instead of paying big bucks for nitrogen fertilizers, I paid pennies for molybdenum. Isn't this the direction graziers want to go in?

In many cases I've found boron and cobalt are also important trace nutrients. The major question is how best to apply these so they are rapidly taken up and incorporated into the biology of the soil. It is not impossible for the cost of application to exceed the investment in these minerals, which is why they are usually included in blends of other inputs.

Make no mistake, soil biology is the nutrient sink that holds, enhances and releases the nutrients of the soil to growing crops. It keeps sulfur from leaching, and amongst other things it fascinates the coy, elusive nitrogen. Those who learned chemistry and think building pasture fertility is a chemical process should go back to school and study biology. While chemical processes work with water and solubility, biological processes hold within cell membranes and organic structures much that would otherwise be lost to water.

Tragically, soluble nutrients tend to concentrate again many miles or hundreds of miles away — and in many cases they overburden and poison rivers, lakes, water tables and discharges into our oceans. Not only does the grower pay handsomely to do this, it is not something a responsible person wants to be the cause of. The main key to accessing most soil minerals is soil fungi, particularly mycorrhizal fungi. Fungi are basically filamentous, and they extend their delicate filaments out long distances through the soil from their centers. Fungi that live in symbiosis with plant roots are called mycorrhizal fungi and they come in two types.

Vesicular arbuscular mycorrhizal (VAM) fungi interpenetrate the fine roots of plants and feed on plant sap in exchange for nutrients. Ectomycorrhyzae colonize the outsides of larger roots and feed on the plant's carbon compounds, but they can be more robust than VAM in spreading their hairlike networks (hyphae) throughout the soil. These symbiotic fungi produce stronger organic acids, such as oxalic acid, than plant roots, which release carbonic acid. Thus they are able to unlock and feed on otherwise inaccessible nutrients in the soil.

This is why in fertilizing soils, graziers should do total mineral tests rather than merely available mineral tests. Though soil fungi can access much that does not show up on an available soil test, they cannot access what is not there. Only transmutation, which is still largely unresearched and debatable, can do that. If the nutrients are there, mycorrhizae are as much as ten or twenty or more times as effective in developing minerals for plant nutrition than bacteria are. Keep in mind that on permanent plantings, such as pastures and forests, that mycorrhizal fungi may develop nutrients from the next field or from even further away. The limit is unknown, but it is known that fungi can cover far more than a football field with their hyphae.

Basically, soil fungi are the developers and reservists of such major cations as calcium, magnesium, potassium and sodium, as well as most micronutrients. But they also access anions such as silica, sulfur, nitrogen and phosphorus.

Biodynamic growers have been a subject of wonder for calling calcium "lime," as well as for their preoccupation with silica. Actually they are more correct to talk about lime. Calcium never occurs in soils by itself. It always occurs as calcium oxide (lime) or its derivatives such as calcium carbonate (limestone) or calcium sulfate (gypsum). Calcium is far too reactive to occur or to be applied as a pure element. No one applies pure calcium; they only apply lime. Likewise with silicon, calcium's opposite polarity. Silicon never occurs in soils as a pure element. It always occurs as silicon dioxide (silica) or its derivatives such as aluminum silicates (clays) or borosilicate (the basis, when fused, of glass — high active plants). Silica is an anion, which means it is negatively charged. In the mineral realm, a balance between positive and negative must occur. In soils silica combines with many other minerals, but its particular ally is aluminum oxide, which it combines with to form clays.

One hears very little about silica in agricultural schools. Why?

First of all, there is little money in silica. Whether you have sand or clay, silica is present. The question is how does it get into plants? Along with lime, silica is more abundant in plants than any other element. In grasses, lilies, bamboo, pineapples, bananas and palms it is the predominant mineral. Often, because there is so much emphasis on enriching soils with silica's opposite, lime, the balance between lime and silica is upset. Then silica does not play as strongly as it should into the above ground processes or photosynthesis, blossoming, fruiting and ripening. Then the digestive, nutritive, nitrogen-fixing processes in the soil escape from the soil and digestion rises up into the canopy so that fungal and insect problems result and the plant or its fruits get digested above ground. Plant leaves and fruits should be hardened and filled with energy by silica, which is closely related to the carbon processes that take place above ground in plants.

This is important because balance is key. Silica is associated with upward growth and the development of nutrition in fodders and foods, and it should always be in balance with lime.

A word of caution is in order here. Commonly growers get the processes of nitrogen fixation related to lime going strongly in the soil and they forget (or have never heard) about the importance of silica. Silica has been ignored for far too long, and diseases and parasitic pests are the most obvious result, even though deterioration of the human will is also related to the decline of silica.

While silica is the most abundant anion in plants and soils, it is not the strongest anion. Along with oxygen, which is what makes silica so anionic, triple bonding nitrogen and its milder sibling, phosphorus, fill most of that role. Chemical industries often supply nitrogen to graziers as urea or as ammonium nitrate. On the pasture these end up oxidizing to the nitrate form and leaching. Altogether the digestion going on in the soil is an oxidative process. As nitrate leaches it tends to take along whatever cations bind with it, such as calcium, magnesium or potassium. The biodynamic oak bark remedy stops this.

In general, biodynamic preparations are a complete nitrogen package to draw every aspect of nitrogen into its role as the moderator of the farm's intelligence. Nitrogen is the key to DNA and RNA. Genes, chromosomes, all genetic material involves nitrogen. Nitrogen carries the master plan, and the nitrogen activity of a farm or ranch is its unique stamp of individuality.

There is no overemphasizing the importance of farms fixing all their nitrogen needs from the atmosphere and recycling this through the farm's own livestock both large and small. That way all of the farm's nitrogen is fully integrated into the organism of that farm. It makes all the difference in the world how a farm gets its nitrogen — whether from crude chemical fertilizers or from its own biological fixation.

In human chemistry nitrogen is associated with consciousness and the ability to think. Any graziers who want to do their own thinking need to fix their own nitrogen. If it seems peculiar that so few people seem to think for themselves, this is because the food they eat does not contain the nutrients they need to be able to do this.

It may seem quirky that biodynamic farms and ranches value their self-sufficiency, and especially their nitrogen harvesting, so greatly. This is not merely a program to avoid buying nitrogen. It is a program to develop the forces of nitrogen and silica so that healthy thought and strong will are developed in those who eat the food from their farms.

With this in mind, applying the biodynamic preparation patterns can do wonders for creating the conditions for that biology to thrive.

While phosphorus is not nearly so electronegative (reactive) as nitrogen, it is able to shift energy states adroitly, and this is the key to many energy transactions in plant and animal biology, to say nothing of soil microorganisms. Phosphorus is the energy bridge in both anabolism and catabolism in plants and animals. Its deficiency is often seen in wine red streaks along stems and leaves, particularly at leaf tips. Nothing unlocks bound phosphorus like mycorrhizal fungi, which particularly like calcium phosphate. This has balanced charge between positive and negative ions and the delicate but powerful fungi love it. Between the way bacteria fix nitrogen and fungi elaborate calcium, phosphorus and other minerals, bacteria and fungi are the two big players in most soils.

Applying phosphorus as phosphoric acid (as in superphosphate) is like taking a blowtorch to the patrons in a bouffant hair salon. The extreme acidic conditions that result from the phosphoric acid, until it binds up with calcium or other cations, melts down to a watery crisp all the delicate fungal hyphae it comes in contact with. Repeated applications of soluble phosphorus can burn up virtually all of the active fungi in the soil. In the short term, there is a burst of available phosphorus that lasts for about three weeks. Then the remainder of what was applied is locked up in the soil, while the means for releasing it has been destroyed.

Remember that it is always better for nutrients to be insoluble but available. This makes rock phosphate, colloidal phosphate, guano, and other unrefined phosphorus sources preferable to soluble phosphates such as monoammonium phosphate (MAP), diammonium phosphate (DAP) and triple sugar (acid phosphate), which are all based on phosphoric acid. Sure, the plants get a blast of phosphorus when soluble phosphorus is applied, but like a junkie on drugs, the initial "high" is followed by a negative reaction that cries for another "fix." Please, do anything else instead of going there.

The same sort of thing is true for nitrogen fertilizers. Nitrogen fixing bacteria in the soil require organic calcium to fix nitrogen. They also require a little molybdenum, as previously mentioned, to manufacture the enzyme, nitrogenase that splits nitrogen out of its powerfully narcissistic love affair with itself— which is why it is an inert gas in the atmosphere.

Enzymes are defined as biocatalysts. As such they trigger reactions but are not used up in those reactions. Thus a very minute amount of molybdenum is necessary to kick off nitrogen fixation. But nitrogen is extremely reactive, as witness the effects of nitrogen based explosives such as gunpowder, nitroglycerine, dynamite, etc. The bacteria that unlock nitrogen from its self-bonding have to have abundant supplies of organic lime (cations) ready to feed into the process to balance the free nitrogen (anions) released. Then the result is rich amino acid production.

One of the more misunderstood processes in agriculture is the role of legumes. Loose talk has it that legumes fix nitrogen. They do not. Legumes are a special class of plants whose relationship with lime and the mineral realm is so intense that they unlock the lime foremost among the most tightly bound minerals in the soil. In the process they drag the blossoming and fruiting process, that usually occurs only at the top of the plant, down into the plant's lower regions. Thus seed formation occurs all up and down the plant and the seeds are enclosed in calcium rich pods. Likewise, leguminous plants form calcium rich shells around their symbiotic roots where they exchange sugars for amino acids. But legumes do not themselves fix nitrogen. Microorganisms fix nitrogen. What legumes supply is organic lime in abundance, and this is the key to the nitrogen fixing process.

Actually in a soil where organic lime is rich and easily available to the azotobacters and azospirilla (clostridia, algae, etc.) that free fix nitrogen, grasses generate more nitrogen fixation than legumes. Grasses, because of their strong silica nature, are better photosynthesizers than legumes. They catch more carbon, make more sugars and provide more carbon compounds as roots exudates to energize the nitrogen-fixing microorganism in the soil, which use up organic calcium as they fix nitrogen. What a balance of legumes does is keep the supply of organic calcium as they fix nitrogen. What a balance of legumes does is keep the supply of organic calcium up. Thus it is appropriate to maintain a balance of grasses and legumes.

One other thing. Aluminum silicates, the anions in clay soils, provide the buoyance that translates into cation exchange capacity (CEC) and a rich storehouse of minerals in the soil. But boron, aluminum's younger, smaller, more reactive sibling, is the element that works with silica to translocate sugars manufactured in leaves to the roots where plants feed their rich carbon harvest to the protein and mineral providing microbes in the soil. Nitrogen fixation not only depends on molybdenum. It also depends on boron, because nitrogen-fixing microorganisms need root exudates, which boron is the key in supplying.

If you have a refractometer, which measures the percentage of dissolved solids in plant juices, take some readings of your forages in the early morning. These readings should be low, because plants should have sent their carbon harvest to the roots at night to feed the microorganisms living in symbiosis with them. High brix readings in the early morning are a sure sign of boron deficiency. Boron works so strongly (as strongly as the explosive nitrogen compared to the merely burnable phosphorus) that only a small amount is needed. But it also is one of the most readily lost elements when chemical fertilizers are used. To get nitrogen fixation working, as it should, keep an eye on boron. This powerful translocater makes energy available to soil microbes and makes nitrogen available to the growing parts of plants.

Speaking of which, nothing so impairs nitrogen fixation as nitrogen salts. Nitrogen salts, such as anhydrous ammonia, urea, or ammonium nitrate are the end waste of nitrogen-fixing microorganisms in the soil. These salts result when soil nitrogen processes go overboard and excess nitrogen occurs as waste. This shuts down nitrogen fixation. When concentrations are high enough, as with heavy nitrogen applications, this kills the nitrogen-fixing microorganisms that do not manage to make spores. Like application of soluble phosphates, which destroys the soil's ability to unlock phosphates, use of soluble nitrogen wrecks the soil's super delicate biological nitrogen fixing processes. As nitrogen-fixing microbes are killed, their nitrogenase is broken down and the key molybdenum is leached from the soil along with boron and other solublized nutrients. Then recovery is a slow process if it occurs at all without remineralizing the soil.

The tragedy of chemical nitrogen fertilization goes beyond the destruction of soil nitrogen fixing processes. When plants take in water, they have no choice but to take up whatever is dissolved in that water. Thus if there is soluble potassium (as from potassium chloride fertilizer salts) they have to take it up, even to the exclusion of such necessary minerals as calcium and magnesium.

No salts are more soluble than nitrogen salts — such as, urea and nitrates. So plants take up soluble nitrogen compounds to the exclusion of the more complex but less soluble amino acids they ought to be nourished with. This makes their cellular chemistry weak and watery. They have to invest considerable energy converting nitrogen salts into amino acids before they can assemble protein chains such as those in chlorophyll. Thus in the protoplasm in their cells there is considerable water and salts, but insufficient proteins. This results in weak and watery plants, susceptible to diseases and insect damage (insects desire simple aminos) with little flavor or keeping quality. Long fiber cotton and fine tasting herbs require that plants take up as much as possible of their nitrogen as amino acids so the assembly of their proteins and DNA proceeds with no interference.

When a plant absorbs abundant nitrogen salts, it looks lush and dark green, but its cell texture will be coarse and watery. It will be a thirsty plant and subject to drought stress. Stock that have to eat it will be in the same shape and will suffer from parasites, diseases and reproductive problems. Moreover, the assembly of chlorophyll in the photosynthetic sites in the leaves will be watered down to the point that instead of 1.5 billion molecules of chlorophyll per chloropast, these plants will only be producing 1 billion molecules of chlorophyll. Correspondingly, photosynthesis will be lessened by a third, even while the plant requires extra energy to convert nitrogen salts into amino acids.

So while nitrogen fertilizers combined with water produce the appearance of lush, dark green growth, keep an eye to cell density, drought resistance, immunity to diseases and pests, and the tender, juicy, exquisite, highly refined flavors of the end products. Nitrogen is the key to quality. And when we produce quality in agriculture we produce food for thought.

As Goethe so aptly pointed out, nature's wellspring in the mineral realm is polarity. In the mineral realm things organize themselves according to their polarities.

But nature's wellspring in the realm of the living is enhancement. Life drives toward ever-greater enhancement. Increasing the life of the land enhances it, and a good grazier, like Abel in the Book of Genesis, is welcome in the eyes of the Creator. Thus the earth can be healed.

Not long ago I joined Malcolm Beck of San Antonio in traveling to west Texas, near El Paso, where George Fore and associates had taken a contract to spread bio-solids from New York on desert acres. Within a year, pasture grasses were springing up as if by magic. Mule deer were heading to the landscape. Nature withheld rain most of the time, and yet the grass flourished. "There are more things in Heaven and Hell than are dreamt of in your philosophy, Horatio" wrote Shakespeare. And there are more things in nature than mere mortals seem able to learn. Beck's observations appear later in this book. Do they answer the problems of the pasture manager? I think they do.

Chapters that seem disjointed and even at odds with each other all come together, I assure you, and as the reader makes a selection for retention and application to his or her operation, the big view Fletcher Sims spoke about will furbish and refurbish the thinking that keeps the grazier on track.

Book One

The Seekers

1

Crossing the Isohyet

"Where chaos begins, classical science stops" so said James Gleick, writing in *Chaos: Making a New Science.* The reason is simple enough. Science likes to deal with exact measurements and shuns disorder as if to preserve its ignorance with hot fertilizers and watered by timer. To listen to urban homeowners, one would think the common dandelion is a menace on par with Attila the Hun.

The grass cover that scientific investigators tend to list outside the purview of their discipline is not grass, but forage. Even the pasture grasslands that once fed millions of buffalo both east and west of the Isohyet Line often has 50 species of forbs, so-called weeds and all known related species.

The Isohyet, much like the latitude and longitude on a Mercator projection, is an imaginary line invented by map makers. Ferdinand von Roemer doesn't mention it in his classic *Roemer's Texas.* Roemer was a German scientist and explorer who became the last trained botanist to see the wealth of Texas in its pristine glory. The Isohyet runs north from the Rio Grande all the way to Canada along the 90th parallel, approximately. It denotes the 30-inch per annum rainfall line. Go west of that line and rainfall decreases at the approximate rate of one inch for every 15 miles. East of the Isohyet rainfall increases at the rate of an added inch every 15 miles per annum. Our yearning for precision suggests a finality that complies with statistics.

Roemer could have told these pioneers of the 1840s that slavery wouldn't fly west of the still unnamed Isohyet. Cotton couldn't make it west of the Isohyet, not with or without slaves. Animals dependent on forage could.

Most plants are extremely sensitive to light the moment they break out of their shell, some even before. Bermuda grass, bluegrass and lettuce are light sensitive even before they break the seed coat, assuming suitable conditions of moisture and temperature. The absence of light determines the rate at which a stem elongates during those early break-out moments, and this makes it important to sink a seed at the optimum depth when planting.

The scenario of a seed come to life contains real life-death drama. Germination takes place in the dark, sending a stem to the surface. At this point elongation is checked by sunlight. At night, in the dark, the rate of elongation increases. This becomes the rhythm. In the presence of high light intensity, stems are regulated to be short and sturdy. If a cloudy pall overhangs the scene for too long, stems stretch out to become long and spindly.

Light continues to figure even after a young plant extends itself above the level of the soil. Obviously a seed can contain only enough food to keep the plant, say, a week or two. After that survival based on seed food storage is difficult because of inadequate food supply. Leaves that are grown in the dark do not expand fully. But if seedlings get adequate light before the food supply in the seed is exhausted, they rush to the task of manufacturing their own sustenance. The process can be dramatic in the extreme, taking only a matter of hours. Once carbohydrate synthesis starts, the plant no longer has to rely on that faltering supply of food in the seed.

Nature and nature's evolution has served up plants for every level of light intensity and duration.

When we tear up nature, we call it development.

Henry Turney once wrote a little book, *Texas Ranges and Pastures the Natural Way*. In it he defined open pasturelands. His definitions were not restricted to Texas, but encompassed the entire United States.

Range is land that was naturally covered with vegetation valuable as forage and is presently capable of supporting such vegetation. Such plant cover consists of grasses, forbs, and shrubs in various combinations and includes prairies, savannahs, and brush land.

It has been recorded in words like these that even a few years ago the atmospheric carbon dioxide was at 270 ppm. As these lines are written, atmospheric CO_2 is over 380 ppm, and is expected to reach 600 ppm even before the end of the first decade of the 21st century. This overload is blamed on fossil fuels.

This much is stated by scientists without any reference to the other side of the equation. Over the centuries, plant growth, pastures included, seem to have gobbled up carbon dioxide the way a funnel cloud gobbles up trailer parks. As we enter the 21st century, we are asked to believe that the good green earth is no longer capable of eating and digesting the carbon overload. Some 16 years ago environmental groups were told that all we needed to do to affect the carbon dioxide fouling the atmosphere was to increase the organic content of our farmland at a rate of 1 percent each year. This would take the CO_2 out of the air and install it back into the soil where it belongs. Left unsaid was how to do this.

These few facts prompted Malcolm Beck of San Antonio's Garden-Ville to exercise the mental acuity for which he has achieved a measure of fame. Working with scientist associates, he concluded that the carbon missing from the soil will equal the excess carbon in the air. The disk plows I mentioned in the foreword merely hinted at the reason for some of this upset. Equally devastating has been the drive to turn pastures into soybean ground for greater protein production.

There was a time, before the frontier was closed in the late 1870s, when the organic content of most soils was at 3 to 8 percent. This load has fallen to 2 percent or less, more often a fraction of a percent. Salt fertilizers and hard nitrogen — especially anhydrous ammonia — have destroyed organic matter and humus with reckless abandon. The resultant problems of soil erosion, water shortages, air pollution and ubiquitous toxicity have caused many environmentalists and farmers to take a new look at the ecology and economies of pasture management, starting with management of carbon dioxide.

Soil and carbon dioxide affect animal and human life on this planet. Worldwide, according to the Beck scientific consortium, human beings account for 8 billion tons of carbon dioxide each year. There are 455 million acres of crop land in the United States. There are 578 million acres of grassland pasture. "If we increased the organic content of crop land alone," said Beck, "this would remove 455 billion tons of what the world generates annually as carbon dioxide."

The implications for grass producers are real and breathtaking. The ranking eco-farmer and organic leader in Texas puts it this way: "When moisture and temperature are correct for plant growth, they are also correct for soil macro- and micro-life to degrade higher life forms." Decay activity in the soil releases an abundance of CO_2, an amount much greater than all the animal life above ground generates and releases, Beck summarized.

"Carbon dioxide is slightly heavier than air," Beck added. "It can remain near the ground and under a plant canopy until needed. Energy from the sun separates the carbon from the oxygen." Carbon is then used to manufacture carbohydrates, which are food and energy for life. Released oxygen then is free to be a catalyst for utilization with food and energy.

This intelligence has been inserted into research work at the Kika de la Garza Research Center in Weslaco, Texas. Joe Bradford, one of the center's researchers, has reported that farmers who use conservation tillage see their soil organic matter content go up the tenth of a percent postulated in earlier paragraphs, this increase confirming itself year after year.

Grazed pastures under the Holistic Resource Management system report a similar organic matter content increase year after year.

Grass producers, meaning beef producers who rely on pastures to feed their livestock, have several inescapable facts they can rely upon. First, the higher the CO_2 concentration in the area of plant growth, that is until the concentration gets 10 times higher than the norm, 2,700 ppm. Greenhouse operators insure a CO_2 abundance by using tree mulch in walkways.

When sunlight becomes too intense, the stomata close, and intake is restricted to the epidermis of plants. Many plants shut down under severe conditions. If photooxygenation occurs, leaves bleach out. The cycle is one of nature's best-kept secrets. In discussing pastures and forage, Malcolm Beck put it this way: "Even though succulent plants take in much water, they still pull a lot of CO_2 out of the air to create carbohydrates they then send out through the roots to feed microorganisms in the rhizosphere. This helps the plant in numerous ways. Up to 80 percent of the carbohydrates a plant manufactures are sent out through the roots to heal the soil. This food source attracts millions of microbes, fungi and soil life that are both beneficial and necessary for forage growth."

This carbon dioxide release into the soil is dissolved into carbonic acid, H_2CO_3, which combines with soil minerals, making them ready for plant uptake.

Transpiration is absolutely necessary if minerals are to be pulled into the plant. In the absence of minerals, transpired moisture is a dead loss. Fully 99 percent of the water a plant pulls from the soil is lost into the air. That is why a high mineral level allows plants to grow faster, even better, with reduced transpiration.

Here is the joker in the deck of cards — when carbon dioxide is high, the plant quickly gets too full and closes its stomata. Moreover, they stay shut longer. This prevents moisture loss to the atmosphere.

Coastal Bermuda grass and other strains of this species that are utilized for forage production in the higher rainfall belts of the state are classified as permanent tame pasture. Ermelo Lovegrass, K.R. Bluestem, Klein grass, Dallas grass, and other introduced species that may or may not be interplanted with clovers and annual grasses are recognized as tame pasture species.

Native pasture includes land managed to utilize stands of native or "escaped" introduced plants for forage where the natural cover was once forest or some other vegetation radically different from that presently growing. Such land includes cut-over forest, abandoned cropland, or former tame pasture that has reverted to native grasses.

Physical characteristics of the land, current economic conditions, or personal desires of the owner might make the conversion of such areas to forest or permanent tame pasture impractical. Control of undesirable brush is usually necessary on such areas, and fertilization and weed control practices may or may not be practical. Controlled utilization of forage is advisable.

Grazable woodland: such areas include forested land managed for timber production which also produces significant amounts of forage that can be moderately grazed without damaging the forest resources. These grazing lands are located in areas where marketable timber or wood products are produced commercially. Grazing should be rigidly controlled on such areas to permit proper tree development and to protect young seedlings from livestock damage.

Controlled grazing of such areas decreases the fire hazard and produces income from forage resources while timber stands are maturing. Forage production is certainly a secondary use and usually decreases as the forest canopy becomes more dense.

The classifications of temporary pasture or rotation pasture areas are cropland planted to forage plants, which are grazed for short periods. Such crops as sudan or sorghum hybrids planted alone or mixed with other forage species may be used entirely for grazing. Grazing may be incidental to crop production as in the case of small grains, or consist of the use of crop residue for forage purposes.

The grazing of such land is a temporary nature usually lasting under one year.

Grass is the natural feed for the bovine. Yet academia would have us believe grains and manufactured feeds best answer the call of husbandry, economics, and high science.

How is this miracle accomplished of changing grains with all their enzyme inhibitors into enzyme-rich foods? The big answer is sprouting, but sprouting is only one cause in a chain of causes that reaches back to creation itself. We look to the laboratory and end up talking about processed foods or esoteric chemistry and lab readouts. And yet it seems we need to look back to our ancestors who didn't think in terms of settled science.

Flocks and herds on American farms during the first half of the last century were often fed sprouted grain. There were texts, of course, but more surprising is the level of sophistication achieved via what might be called a folklore system. The percentage of nonstructured carbohydrates, sugars and starches, that conspired to deliver the most favorable nutrition were worked out by trial and error. Sugars are generally thought of as sucrose, fructose, glucose, lactose, dextrose all extracted from feed at 39 C, stirring for one hour. Starches did not yield all their sugars, not in the crude laboratory method described above.

It was the sprout that unlocked and changed the sugars. And this code of sugars interdicts health problems in animal husbandry. Release of sugars has a counter balancing effect. It raises the enzyme level. Now the grain becomes more digestible, and these enzymes help in the digestion of other feeds.

It has been observed that only birds are naturally designed to consume ripened grain whether harvested in the field or fed from the bin. These grains are loaded with phytates, better known as enzyme inhibitors. Merely soaking grains has proved most beneficial when feeding grain in animal husbandry. At a bare minimum, this soak nullifies the phytates. Moreover, the quality of the protein is improved by being made less soluble.

Following this line of thought, either observation or teaching reveals a high solubility of protein in forages. Schoolmen tell the farmer that because of a soluble protein intake, the process orders protein bypass, a purchased input and therefore a subtraction from the bottom line.

Sprouted grains, when fed, erase that cost factor whenever grains are fed.

Sugars are not confined to grasses just as pastures are not restricted to grass. Most pastures have 40 to 50 plant species living as part of soil's green canopy. Many weeds and forbs are 50 percent sugar. The term weed is a fiction agreed upon as a monument to human ignorance. "Nature does a very good job producing high quality protein," said Jerry Brunetti at an Acres U.S.A. seminar. His drought scanner tests discovered quality and a bypass fraction high in minerals and highly structured carbohydrates.

We are appreciative of plants with medicinal values. Yet our main concern is nutrition, and the reality returns us to grass from sprouted grain and its unique capacity for uptaking available minerals. Chemical tradition says "as ions," and some organic folks agree for "whole molecules."

We have about 1,180 million acres of farmland in the United States, improved and unimproved. Go west of the Isohyet and you have subnormal rainfalls, mostly grazing land in the west. Half of these acres end up as complete wilderness except for livestock. There are places where it takes 20 to 35 acres of grazing land for a mouth full of grass to feed a cow and raise a calf.

"The livestock as a lab processes all that natural grass production." There are about 75 million acres of hay. This is used to carry cattle through two or three winter months. It takes livestock to process this hay. I think this is as it should be. But now we have cows processing 80 percent of the corn crop. Linseed, cottonseed and soybean meal are being processed by cattle. It is merely one position between raw materials and earned national income. If we lose that production or don't pay for it, then look out for debt expansion.

When we speak of grass we usually mean forage, that mix of forbs and grass—tame and wild—that gives each paddock or open range its character. Henry Turney, that savvy cowman who brought pasture course study to the University of Texas system, held the opinion that the best grass was the one that nature and evolution provided. Roemer would have agreed. This suggests that an examination of the anatomy of grass is in order.

In writing, assembling, quoting, abstracting and recapturing the values of grass, I have relied on countless interviews, statements, book quotations and unpublished manuscripts. I have found that real farmers developed their own system, usually as a sub or temporary project. Ted Slanker of Powderly, Texas, owner, entrepreneur, a mature leader in the grass-fed industry, did. He not only raises grass-fed cows and bulls, he crafts some of the really hard hitting statements making the case for grass-fed, facing down the feedlot product as a disgrace to a civilized society.

Ted Slanker writes a controversial column, "Just Managing to Get By," the emphasis being management as the key to prosperity unlimited. His emphasis is on management of cow-calf operations without hay and feed supplements using rotational grazing, developing salad-mix pastures and all the topics that relate to keeping grass as the basic input. He specializes in the planting and growing of pastures and how everything comes together during a 45-day breeding season. He handles cattle via runways and paddocks and uses dogs and horses in his type of operation. Considering that all species of domestic livestock graze and/or browse, good pastures are essential for economic and animal-friendly livestock production. More important, when pastures and browse are our livestock's primary "feed," only then will animal products appropriately meet man's nutritional requirements. When livestock are fed grain and other unnatural feeds, they suffer nutritional deficiencies. Consequently, food products produced by or from them are detrimental to the health of all who eat them. Slanker writes:

This view about animals and grain seems preposterous to most folks, yet it makes sense when you think about it. And backing up simple common sense is peer-reviewed scientific findings that have been pouring forth for years now from many of the state-of-the-art nutrition laboratories at major universities located all over the world. To learn more, do a search for "omega-3 fatty acids" or "CLA" on the Internet. The quantity of information will amaze you. Then do a search for "grass-fed meats."

When folks in livestock production see a lead-in like the one I've just used, the vast majorities are convinced I'm way off the page. But they're so brainwashed by big business interests, behemoths who manufacture most of our nation's food, that they don't have a clue about what's factual in nutrition. As a result, they don't know the difference between real food and fabricated food and couldn't care less. You can tell who they are. Most of them, plus their family members and friends, are overweight and struggling with ailments such as cancer, diabetes, heart disease, allergies, arthritis, mental disorders, and other diseases associated with failing bodies. They're the medical and drug industries' best customers.

The findings of modern, peer-reviewed nutritional scientists are readily available for all who seek them. But the power of consensus reality affects the thinking processes of most Americans. That's right. Most Americans are incapable of independent thought. That's why there are thousands of university professors specializing in livestock production and meat science who refuse to give so much as a nod to the findings of the nutritional scientists employed by their same universities!

We know that plankton and algae are the foundation foods for all animals in the sea. And the role leafy-green underwater plants play in the sea is the same for leafy-green plants (grasses, brush, and trees on land). Therefore, because animals cannot "manufacture" many of the various essential fatty acids required for proper body function, they get them from their food. That's why the fatty acid profiles of leafy green plants have dictated the overall fatty acid profiles of animals in the sea and on land for many millions of years. With that information even a halfwit can understand why deviations from the fundamental feeds (leafy-green plants) cause the bodies of all animals to fail in time.

Obviously, pastures are very important for not only the health of our domestic animals but also for man if he intends to eat food products derived from his livestock. What's really neat about this indisputably normal approach to food production is that big business does not own the land. Individuals own most of the highly productive land in America. Therefore, the power of real-food production is outside the grasp of big business monopolies and rests solely in the hands of the "little people."

Pastures empower landholders, but only if they recognize the opportunity. Most won't, but after they pass away maybe their children will figure it out. On the other hand, for those of us who are quicker studies, if pastures are power, are we focusing on optimizing our leafy-green plant production?

Forages take root in soil, from which they derive a relatively and surprisingly small percentage (from 0.5% to 6%) of their nutrients. Most of their nutrients come from air and water. But when it comes to economical forage production, it's the soils we should examine first because most often the limiting factor is in the soil makeup.

Throughout the world all soils near the surface are affected by minerals that migrate up from the molten depths of the earth's core. That's why soils differ not only from ranch to ranch but from field to field. That's also why gold mines, copper mines, zinc mines, lime deposits, etc., are usually located in different regions of the country. In addition, soil productivity is also impacted by latitude and whether or not the weather is wet or dry, hot or cold.

There are 17 essential elements that greatly impact all plants. They are carbon (C), hydrogen (H), oxygen (0), nitrogen (N), phosphorus (P), potassium (K), calcium (Ca), magnesium (Mg), sulfur (S), iron (Fe), manganese (Mn), boron (B), molybdenum (Mo), copper (Cu), zinc (Zn), chlorine (Cl), and cobalt (Co). There are additional elements that play minor roles in plant development. Not only does the quantity of these elements vary depending on soil types, but the availability of the elements to the plants varies with soil pH and/or the balance of the elements in the soils. This is why ranchers should learn everything they can about their soil types, capabilities, and pH.

Only after soils are understood in at least a basic sense can a livestock grazer effectively focus on forage development and management. Naturally every grazer wants to optimize his forage production. He does it by determining his limiting factors and addressing them directly with fertilizers/lime and/or by planting pasture plants that thrive best in his ecosystem. This involves some research, arithmetic, and faith in science.

Only after the limiting factors have been addressed can we select the pasture plants. Let's say my main objective is to grow abundant forages 365 days a year. Of course, that may be an impossible objective, but it's my aim. I want winter and summer forages, perennial and annual forages, and about an even mix of grasses and legumes.

To start with, a grazer must carefully study the growing periods for various forages in his region. In my area fescue, forage rye, and ryegrass do a terrific job in the winter and Bermuda grass, Johnson grass, and crabgrass do a terrific job in the summer. Many additional winter and summer forages work as well and I've got lots of them.

I want a mix of perennial and annual grasses for two reasons. Perennial grasses make a sod that in the winter helps support the livestock and protect the soils when they are saturated with water. In both summer and winter, annual plants provide more protein and are more digestible. Therefore, livestock get a bigger bang for the buck from grazing annuals.

A necessary component in my pastures is legumes. Legumes provide a nutritious feed for the livestock, and in most cases legumes will increase nitrogen availability for the grasses. With an excellent legume/grass mix (50/50) the legumes may fix up to 100 pounds of actual N per acre.

An input of 100 pounds of actual N per acre is a very valuable contribution from the legumes given the availability of other limiting factors; an extra 100 pounds of actual N can increase forage production by 4,000 pounds per acre. If I bought ammonium nitrate fertilizer to supply 100 pounds of N, I'd have to apply 300 pounds of fertilizer per acre. At $225 per ton, that would cost me $33.75 per acre.

Compare the cost of fertilizer to my last year's legume/annual grass mix. I used an old conventional 15-foot seed drill. Nothing fancy. I applied five pounds of ryegrass, one pound of crimson clover, 1.25 pounds of arrow leaf clover, 0.25 pounds of white clover, two pounds of peas, and one pound of vetch per acre. The cost of the seed was about $7 per acre. Using the conventional seed count per pound, I planted 45 seeds per square foot or one seed per 1.75 square inches. The result was fields of clover so dense in many areas I couldn't take a step without stepping on a legume. In other areas where the soils were not so friendly, the results were not as gratifying. I anticipated this by selecting a mix of legumes with differing capabilities. So where some couldn't make it, others still grew. What I like most about legumes versus the fertilizer buggy is that plants reproduce. In the areas where they thrived, they'll come back like gangbusters.

I take additional steps to augment my forage production. If I must clip a pasture I'll do it in the winter or at other times when the grass is very short. I do not like to cut grass. I will not cut hay. I want the cattle to harvest the grass and recirculate the nutrients. In that regard, I am really big on building organic matter in the soils and view hay cutting as the pasture man's worst enemy. Organic matter protects soils from erosion and the decaying process produces more N. Look down at the ground in your pastures. You should not be able to see soil, only litter.

Another critical component of pasture management is controlled grazing. Livestock do not understand the private property limitations of your ranch. Cattle are instinctively programmed to eat and move on. That's why grazers must serve the nomadic habits of their livestock. Four pastures per ranch is the absolute minimum. To optimize forage growth and utilization one needs a minimum of 12 pastures per livestock grouping. Only with controlled grazing can one optimize forage production and utilization while raising more pounds of beef per acre. Additionally, by managing my grazing periods, I optimize the reseeding of valuable legumes and other annuals.

If you're in the livestock business, pasture management is one of your more important studies. Not only can it lead to greater profits, but it will permit you to raise the healthiest livestock food products in America. Hearty pastures are the grazier's best friend.

Thus writes Slanker. Many people do not have respect for Slanker's columns because in their opinion there is no difference between feedlot beef and the Omega-3 rich meat protein harvested from well-maintained pastures. Feedlot bloat, *E. coli* in the hamburger, and even the threat of Mad Cow disease and para-T (Johne's disease) and a prelude to pandemics does not concern editors whose knowledge of epizootics is no more advanced than it was in the Stone Age. Here are a few of the reasons Slanker gives to explain why the mainstream livestock industry rejects the simple and obvious pasture system discussed in this book.

1. Packers, who have expensive plants located too far from the best grass-growing regions of the country;

2. Feeders, who own feedlots that currently feed all of the nation's young cattle that go to slaughter;

3. Stocker operations, thousands of whom purchase calves to put on winter wheat;

4. Grain growers, who have large investments in machinery and facilities to raise and ship huge volumes of product to the feedlots and cow-calf producers;

5. Feed mills and feed store, which are dependent on cattlemen supplementing their cattle;

6. Drug companies, which may experience a drop in sales if cattle are healthier on grass than they are in the feedlots (they will certainly lose a lot of hormone implant business if producers leave their bulls intact and market them as all natural, grass-fed beef);

7. Our nations' health industry, which is larger and more costly than it's ever been before and which treats primarily health problems related to diets heavy in bad fats (better diets mean fewer health problems and less business);

8. Food processors and manufacturers, which will face embarrassment and lost sales as new nutritional guidelines kill off previously successful products;

9. Food retailers, which will be reluctant at first to switch to, or even introduce, new products for fear of losing business on mainstream products;

10. Projects sponsored by, and employees of, the Beef Check-off, which will see a drastic reduction in funding if most calves are retained rather than sold (additional, the Beef Check-off will find it nearly impossible to market grain-fed and grass-fed beef at the same time, so it will try to stick with what works for now);

11. Sale barns, which will experience a major reduction in the number of cattle marketed if producers switch to marketing direct through meat marketing alliances;

12. Producers, who can be the main beneficiaries of the grass-fed management and marketing approach yet will still ridicule and stonewall grass-fed beef because the entire approach will be foreign to them and they just don't like change; and

13. Educators, who have been teaching old practices but will feel foolish teaching new practices they once ridiculed (also, when educators start to teach new grazing and grass-fed approaches they

will be turning their backs on old friends who own or are employed by businesses that will have to downsize because of the grass-fed beef raising and marketing approach).

John Ingalls called grass the forgiveness of nature, her constant benediction. Here I call it the industry of grass.

In taking an extended look at pastures, I have visited and revisited the vast miscellany of operators who have cast their shadows over the pages of *Acres U.S.A.* for 35 years.

Their reports in print and over the rail fences that mix definitions with barbed wire tell more than most texts about what is happening in and on pastures. Cattle pens and grass make a strange mixture, and yet the feedlots decide how nature's bounty is to be used.

Slanker became a serious Polled-Hereford breeder in 1974, raising bulls and heifers without supplements except minerals and salt. The common denominator is grass and a set of settled findings that ought to be framed in any farm office for consistent review. These are presented here because they define both what is right and what is wrong with ranching, the holy process of turning grass into beef, all while maintaining a bottom line. He teaches widely accepted management practices with the icy professionalism of a surgeon removing cancer. Moreover, he holds up the mirror of self-sufficiency so errant Don Quixotes can see the folly of the industrial model called feedlots.

The Omega-3 benefits of grass-fed beef have become current coin during the past dozen years, and yet resistance remains mobile, much like Bram Stoker's undead.

The classification and management of such land is based on the assumptions that, in light of present knowledge, the most desirable cover is similar to the climax cover. The proper management of such land consists primarily of regulating the degree and time of use made of forage by grazing animals. In addition to such grazing management, practices such a brush control, additional fencing, range seeding, adequate water facilities, and other improvements may increase forage production and economic returns from such land.

Permanent grazing land lying in the western half of Texas, excluding irrigated areas, is a typical example of range. This is the arena in which the seekers ask questions, then stay on for answers.

2

The Anatomy of Grass

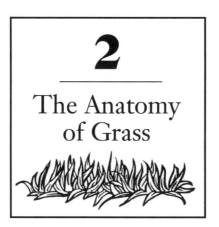

Wise old graziers who manage to keep a disease-free herd year after year have a truism: "Worse than overgrazing is not grazing at all." Reclamation absolutely depends on it. So does maintenance. The methods and objectives of holistic resource management are covered elsewhere in this book. For now it is enough to point out that grazing for health requires enough grass leaves remaining after a grazing sweep to enable photosynthesis. Urine and manure gifted the soil by tight herds, most of it stomped into the soil, the carbon dioxide flush is more than ample for rapid regrowth. This CO_2 flush stays in the canopy of the grass to feed the pores. Saving soil moisture is dependent on the stomata valve being closed before all the soil moisture is transpired into the air.

The marijuana grower with a clandestine indoor operation often relies on fertilizing the air. Accordingly, he pipes CO_2 into the growing chamber to achieve lush growth. The growth arrives, sure enough, but the quality suffers.

> The word *grass* supposedly evolved from an old Aryan root, *ghra-*, to grow. It is related to "grain," "green," "grow," and the Latin *gramen*, grass.
>
> The Oxford Dictionary gives the primary definition of *grass* as "herbage in general, the blades or leaves and stalks

of which are eaten by horses, cattle, sheep, etc." This elemental usage is reflected, for example, in the Bible (". . . all flesh is as grass, and all the glory of man as the flower of grass"). Now, however, grass primarily refers to the natural botanical family of grasses *(Gramineae* or *Poaceae).* Grasses belong to the seed plant subkingdom *(Spermatophyta)* and thereunder, 1. to the subdivision of angiosperms *(Angiospermae)* with rudimentary seeds (ovules) enclosed in an ovary, and 2. to the class monocotyledons, the embryos of which have one seed leaf, or cotyledon.

The *1948 Yearbook of Agriculture* thus defined grass, and for several hundred pages examined grasses from Main to Florida, California to Washington, and all parts in between. Unfortunately, these descriptions do not come to terms with the real anatomy of grass or explain how and why grass uptakes more nutrients than all other crops and why it is absolutely necessary for herd health.

When we say "grass," we usually mean forage. Others use the term as slang to identify cannabis. The golfer won't call a putting green "grass" — it's a "green," period! John J. Ingalls, an early seeker, gave a speech that was a paean to bluegrass, and when Ann Wigmore starting juicing grass, the anatomy of grass took on new dimensions. It is a function of literacy for a reader to comprehend the meaning of a word according to the context of its usage. Much like the average farmer, I thought I understood grass, this after a dozen years in farm journalism. But that was before I met Dr. Charles Schnabel. He presented for publication a short article entitled "An Ecological Sputnik." The title denotes the era. Schnabel died a year or two later, but he gifted me his research findings, some of which are quoted below.

"Ecology deals with the unusual relations of an organism with its environment," Schnabel wrote, "this as a prelude to an autopsy on grass that has most cowmen shaking their heads. To understand all the relations of an organism, including maintenance with its present environment, we must follow ecological clues clear back to 250 million years ago. Man's survival depends on finding out what started and stopped the explosive planet growth which made the fossil fuels possible." Almost all ecologists agree that planet Earth must have supported a thousand times more plant growth than it does today. Coal beds and oil deposits didn't just happen. All fossil fuels have one thing in common. They are a consequence of reducing conditions. Those same conditions prevail today in water-logged soils which have produced paddy rice for a thousand years without the addition of nitrogen.

Paddy rice soils get their nitrogen from blue-green algae, the algae that forms green rice oil. In other words, biological nitrogen fixation is a purely reductive process. The plan here is that reductive conditions should be maintained around the roots of farm crops, especially grass.

Schnabel tried his theory on rye, one of the cereal grasses. He grew his crop on summit silt using simple extraction. This revealed a possible production record of 21.78 tons of dry grain per acre. This rye plant was grown with seaweed.

During a recent tour of the Kika de la Garza Ag Experiment Station near Westlaco, Texas, an agronomist lionized certain fertilizers but pronounced coal nearly worthless. That's not what Charlie Schnabel's tests indicated. He found coal and oil shale worth more as fertilizer than as fuel. As Dr. Carl Oppenheimer of Austin, Texas, has demonstrated with his blend of RNA bacteria, coal and even crankcase oil can become a source of useable carbon after these microorganisms are done with it. Investigators often create more questions than answers. What were the variables that governed record production at one spot and a completely different result a foot away? The flip answers are always available — not so easily identified are the variables that dance before our eyes.

Farmers are always on the hunt for greens that make the connection between chlorophyll and gain. They turn to alfalfa, a time-honored forage. Yet a morsel of alfalfa meal from 5 percent to 20 percent of the poultry ration ignores the kidneys. There is always a but! The principle of alfalfa is in the leaves. Accordingly a 15 percent protein crop can be fed at higher levels than alfalfa leaves with 30 percent protein.

These few asides are presented here to call into question the idea that anything green is a gift from heaven. As a matter of fact, all the vegetables have been tested, not only by Dr. Firman Bear, but by hundreds of researchers. None improved the record of alfalfa. Spinach, mustard, turnips, collard greens, and two varieties of lettuce have proved no more effective in regeneration than their respective ashes. Alfalfa proved twice as effective.

Simply stated, all chlorophylls are not the same — whether flat or saponins, etc., they instill biological differences in the plants so dazzlingly we are required to note in the expertise of the greatest nutritional expert on planet Earth, the cow.

It was serendipity that gave early researchers the clue that kicked open the door to the chlorophyll-vitality connection, when immature wheat and oat grass chop was accidentally fed to poultry. The immature grass-fed birds averaged 94 percent production while control birds reduced production from 45 to 32 percent during the test months. The test hens remained free from degenerative diseases.

Grass, not corn chop or silage or protein bypass, accounts for bovine health, and the grass-fed cow knows it. Even the friendly dog goes to grass when its blood needs rebuilding.

These are rules that ought to be posted on the shaving mirrors of cowmen, and cowwomen, if you will:

1. Dehydrated grasses must contain 30 percent or more protein. Lower quality grass will debilitate the animal.

2. Grasses must be cut just before they joint. Proteins and vitamins in grass peak immediately before jointing and fall rapidly after jointing.

3. Grasses must be quickly and carefully dried in order to preserve vitamins and color.

4. High protein grasses must comprise 20 percent of the ration.

5. The ration must not contain more than 3 percent meat scrap.

6. Livers previously damaged by meat scraps do not recover.

Are poultry tests useful when discussing grass and pastures? They are if we are to understand the benefits and some of the shortfalls in pasture management. Simply turning poultry out in pasture is not adequate because the grasses are at the proper stage too short a time and perennial grasses hardly ever contain 20 percent protein. Even when the grass is available, few hens consume the required eight pounds of fresh grass per month on a range.

Pastures are generally populated by various species of perennials, some of which retain their reserves in the grass's joint at different times. The bovine, as nature's finest nutritionist, sorts out the bits best suited to maintain health and milk flow, always choosing the best unjointed grass available for its lower row of teeth.

There was a time when the bison was present in vast herds from the Great Lakes to northern New Mexico. The buffalo ate only grass and in captivity generally refused alfalfa hay. The horse, sheep, and grazing wildlife thrive indefinitely on quality fresh grass.

All these animals seem to realize the value of young grass. "They have always eaten all of it as stupidity would permit." Charles Schnabel told this journalist. "Otherwise we would have come to biological and economic destruction long ago."

There is irony in this. Agronomists search the globe for better crops and have untapped a full measure of the universal crop, grass. Thus all the cereals are at the bloom stage when they have lost 50 percent of their biological values.

In *Unforgiven*, I pointed out that the unpaid work force of an animal population alone is capable of harvesting all the untilled acres, always using its intelligence to harvest at exactly the right time if given a chance.

Dr. Schnabel's Findings

The amount of grass laying hens will eat when it is cut fresh and fed to them at daylight is a fair measure of its quality. A hundred hens will eat about one pound of fresh grass of each percent of protein in the grass on the dry basis. In fact, this was the way a secret of grass was discovered.

A flock of Leghorn hens fed all the fresh 30 to 40 percent protein grass they would eat as a supplement to their regular mash-and-grain ration averaged 89 percent production from April through August. On 14 days during this time they laid over 100 percent, the peak day being 126 eggs from 106 hens on August 3. The check pen, getting the same mash and grain ration, but with alfalfa as the source of greens, averaged only 40 percent production during the same period, and their mortality was ten times that of the grass-fed hens.

The yellow pigment never faded from the beaks or legs of the grass-fed hens, some of which were killed for table use just to see what they looked like on the inside. The most striking thing was the appearance of their livers, which were a dark mahogany color, and the surface of the livers glistened like a mirror. The alfalfa fed hens had light tan-colored livers throughout the test.

The grass-fed hens produced a dozen eggs on three pounds of total food, while the alfalfa group required seven pounds of feed to produce a dozen eggs. If we figure the mash-grain part of the ration at 3 cents a pound, the eggs at 30 cents a dozen (1950 prices), the grass (on the dry basis) was worth 6 cents a pound for the food it saved, and another 15 cents a pound for the increased production, to say nothing of the greatly reduced mortality rate. The grass left a residual effect on the hens for six months after the feeding was stopped.

All the grass-fed hens were held over for another year, but 80 percent of the alfalfa fed hens were culled out at the end of the test as worthless for future production. The following spring the grass-fed hens were placed in batteries (five hens to a coop, 30 by 30 inches square) and the grass feeding continued at various levels. One coop of five hens laid 23 eggs in three consecutive days and averaged 102 for the month of May. Alfalfa-fed hens killed each other when five were confined to such a small space for any length of time.

The facts revealed in this test can be checked by anyone who wants to repeat the experiment. The mash and grain was fed in hoppers, and the hens ate three pounds of grain to one of mash. The grain part of the ration was a mixture, mostly wheat. (Be sure of the quality of your grass and feed it at daylight. If the hens fill up on grain and mash, they won't eat enough grass.)

For five successive years Schnabel took grass samples all over the United States, following the seasons so as to catch cereal grasses at the jointing stage. The highest-quality grass ever found was oats (45 percent protein) growing in an old feedlot in Missouri. The lowest quality ever found (4.5 percent protein) was also oats, intended for use in a cattle-grazing experiment. Is there any wonder there should be conflicting evidence in the literature about the feeding value of grass when a difference of 1,000 percent in quality is possible?

Cereal grass which tests 45 percent protein at the jointing stage may test only 10 percent at the bloom stage, but grass that tests only 4.5 percent protein at the jointing stage is straw at *any* stage of growth.

The best and poorest samples were taken a thousand miles apart and in different years, but no observer could ever forget the contrast between them. The good oats grew to a height of 18 inches before jointing, and some seed culms had as many as 50 tillers. The 4.5 percent protein oats were jointing at the height of 6 inches and did not average one tiller per seed culm.

The good oats would have yielded over a ton of dry matter per acre and would have made several cuttings. It is possible for one planting of annual grass to make ten cuttings if it is cut every time it reaches the jointing stage and the soil has the "kick" to keep it coming. The poor oats would have yielded about 200 pounds of dry matter per acre and one cutting would have killed the stand. They were a sickly yellowish-green color and the leaves were rough and brittle to the touch. They had the typical corkscrew appearance of poor grass.

The farmer who grew the 45 percent protein oats was not aware he had done anything unusual. It was later estimated that one acre put 500 pounds of gain on four medium-weight steers in 30 days. The oats were planted March 1, grazing was started March 20, and was stopped April 20, because the steers could not keep up with the growth and most of the feed had jointed. Judging by what the field would have yielded before grazing, and allowing for growth during grazing, these four steers could not have eaten more than 2,500 pounds of dry matter. Thus, it took only five pounds of dry grass to make a pound of gain, and the steers topped the market as grain-fed beef although they had not eaten a pound of grain.

Incidentally, good grass is responsible for the worldwide reputation of Kansas City steak, but corn usually gets the credit. Farmers in the Kansas City area graze more cattle on cereal grasses in the spring and fall than in any other area in the world. A calf would die within a year on a sole diet of corn and milk.

According to drylot feeding standards reported by USDA, it takes 12 pounds of grain to produce a pound of gain on medium-weight steers. That means this acre of 45 percent protein grass saved 6,000 pounds of grain, which at a cent a pound was worth $60, within 50 days from the date of planting. This is more grain than the same acre would produce in two years without any grazing.

The sequel to this story is that the same acre was in oats again three years later, following a crop of potatoes and a crop of corn for grain, and the farmer wondered why the same acre made only 200 pounds of beef under similar conditions of grazing. The reason was that the quality of the second oat crop had dropped to 25 percent protein, but even then it was worth $24 for 30 days of grazing. It cost just as much to plant the 4.5 percent protein oats as it did the 45 percent protein oats. The moral should be clear: On the same basis, 10 percent protein grass is worth about $2 per acre for grazing purposes.

Any farmer would consider a return of $60 an acre for 30 days of grazing a very profitable operation (at 1950 prices), but in reality grazing 30 to 45 percent protein grass is a wasteful practice. Grazing animals utilize only about 10 percent of that kind of grass as a sole diet. The calories of a cattle ration can just as well come from a much cheaper food, even straw, when they have just a little good grass. If that acre had been preserved at the peak of its quality and fed the year round at a 1 percent supplement to ordinary cattle rations, it would have saved 60,000 pounds of grain, worth $600 at a cent a pound. That is more grain than the same acre would produce in 12 to 20 years without any grazing.

Every acre of 40 percent protein grass will save 100 bushels of grain if grazed for 30 days and will still make a better grain crop than 20 percent protein grass without any grazing. Thirty million acres of 40 percent protein grass (10 percent of the cropland in the United States) would cut the feed cost of producing meat, milk and eggs in half. A farmer could well afford to spend $100 an acre to make 10 percent of his cropland grow 40 percent protein grass for grazing purposes.

What Is a Protective Food?

It is well known that grazing animals can live on grass alone, and pretty poor grass at that. It has been assumed that herbivorous animals could live on *any* of the common leafy green crops, but this is not the case. A guinea pig is herbivorous, and yet it will die in eight to 12 weeks on a diet of head lettuce, cabbage or carrots, and will grow at only half its normal rate on a sole diet of spinach. But a guinea pig thrives on a solid diet of *grass*. A super race of guinea pigs was developed in five generations on a sole diet of 20 percent protein dehydrated grass.

When the first vitamins were discovered, leafy green vegetables were placed at the top of the list of so-called protective foods, and there is no doubt that all green leaves are rich in the vitamins associated with photosynthesis, the same as green grass. Then why won't any of the common leafy green foods support normal growth and reproduction in the guinea pig? Is it because they lack something, or is it because they contain some harmful substances?

Neither man nor animal dares eat the leaves of more than 15 flowering plant families, out of a total of 332. Most of the greens used for human consumption are supplied by only three families. The *mustard family* supplies Brussels sprouts, broccoli, cabbage, kale, kohlrabi, mustard, radishes, rape and turnips. The *composite family* supplies dandelions, endive and lettuce. The *goosefoot family* supplies beets, chard and spinach. Every one of these leafy green vegetables contains more of one poison or another than the government would allow a food manufacturer to put in processed foods.

It is well known that many of the wild species of the mustard family are poisonous to livestock. *The poisonous principal is mustard oil,* which in large doses causes chronic enteritis, hemorrhagic diarrhea, colic, abortion, nephritis, harmaturia, apathy, and paralysis of the heart and respiration. In small doses over a period of time it damages the thyroid in some way yet unknown. When fed to laying hens, any specie of the mustard family will cause green-yolked eggs — wrongly called "grass eggs" (grass never causes green-yolked eggs).

Lettuce contains an amaroid, lactucin; an alkaloid, hyocyamine; an opiate; and rubber. *The harmful substance in spinach is oxalic acid.* Too much spinach in a poultry ration causes soft-shelled eggs.

In order to get the optimum supply of vitamin A and vitamin C, the average person must eat nine ounces of leafy green foods per day. Certainly, no food should be called protective which will not pass the guinea pig sole diet test.

It is well known that alfalfa cannot serve as a sole diet for any of our domestic animals. J. Sotala found the protein of alfalfa to have a value of 51, as compared to 94 for corn-silage protein. Morrison and Maynard Turk found alfalfa protein to have a value of 50 when used as a sole source of protein for lambs. The value apparently rose to 72 when sugar or starch was added to the alfalfa diet. This looks like the dilution of something rather than a lack of something in alfalfa; 20 percent alfalfa in a poultry ration causes serious kidney damage.

Numerous experiments show that grass pasture alone can support a milk flow of 40 pounds per cow per day, but 40 percent protein grass as a sole diet will support a milk flow of 80 pounds. Furthermore, a pound of (dry) 40 percent protein grass will make four pounds of milk, while a pound of alfalfa makes only one pound of milk.

The most surprising thing is that a cow could thrive indefinitely on a sole diet of 40 percent protein grass, yet any farmer knows what would happen to a cow on a sole diet of any concentrate containing 40 percent protein.

Grass protein seems to be unique in that it can serve as a source of energy without injuring the liver or kidneys, although this is a wasteful use to make of it. Some new definitions of protein quality are long overdue.

In a USDA bulletin, Graves, Dawson and Kepand report that every time alfalfa was substituted for grass pasture there was a sharp drop in milk production. Others report very low milk production on alfalfa alone. Whenever alfalfa pasture contains considerable grass, milk production is usually satisfactory in the spring because cows always graze grass before the jointing stage when given the opportunity.

Grass & Liver Function

The most notable effect of good grass is the beneficial change it causes in the liver. Not one of 20 leafy green vegetables would cause those dark mahogany colored livers which are so conspicuous in grass-fed hens. Furthermore, grass-fed hens will not approach 100 percent production until these liver changes take place, but when liver injury has gone far, even good grass won't change or correct the damage.

Unfortunately, none of the tests for liver function show any sign of liver failure until 90 percent of the liver has been destroyed, and by that time the victim, whether animal or man, has died of some other degenerative disease, or the liver damage is beyond repair.

The world has everything but good grass. The ancient shepherd apparently knew that "all flesh is grass," but modern man has had to relearn this homely truth the hard way. The liver changes caused by good grass are too obvious not to have some connection with the prevention of degenerative diseases.

Dr. F.M. Chichester makes the best summary of the situation from a medical point of view: "In the early stages of vitamin deficiency, the thyroid gland and other endocrine glands are overactive. This overactivity causes the body to lose enormous amounts of calcium, iodine and iron, which leads to goiter, anemia, nerve degeneration, diabetes, paralysis of the limbs and gastric ulcers. Finally, in the later stages of vitamin deficiency, when the glands have become exhausted, sterols accumulate in the body and form gall stones, cataracts, hardening of the arteries and the most malignant form of cancers. The vitamins of natural foods are best because they have no chemical imbalances."

Dr. Chichester's theory explains the universal prevalence of tooth decay, even among people getting twice the supposed minimum requirement of calcium. A small amount of good grass in the human diet prevents tooth decay, which is the result of other degenerative changes in the body. Don't ever forget that degenerative diseases start their "spiral of destruction" before a child is born. Stillbirths are certainly not a *normal function* of motherhood.

It is folly to dose ourselves with one or two vitamins when we know nothing about their relationship to 50 other food factors. For example, it takes 20 percent of 20 percent protein grass to serve as the sole source of vitamins for a guinea pig, but 5 percent of grass will fully replace any one vitamin which is purposely left out of a guinea pig's diet. This can only mean that some of the vitamins must be interchangeable.

Grass does many things in animal nutrition which cannot be accounted for by its known vitamin content. For example, 40 percent protein grass is only four to five times as high in known vitamins as 20 percent protein alfalfa, yet 2 percent of 40 percent protein grass will perform miracles in a poultry ration as compared to 10 percent of 20 percent alfalfa.

Either God or man is still mixed up on the subject of vitamins. For example, only man, monkey and the guinea pig are supposed to need vitamin C. Good grass is the richest natural source of both vitamin A and vitamin C, yet grass has been considered fit only for cow feed, and grazing animals are less than 1 percent efficient in transferring these vitamins to meat, milk and eggs.

Vitamin C and good-quality protein are the most expensive food factors in the human diet. How could it be otherwise when the world's richest source of these factors is a total loss as far as human nutrition is concerned?

The Challenge

Surely, if 45 percent protein grass has been produced accidentally, American farmers can grow 40 percent protein grass on purpose. If it pays — and all the evidence indicates that it does not pay to grow any other kind — high-protein grass is truly green gold, and the pay dirt is the top 6 inches of good earth.

If cereal grasses are cut each time they reach the jointing stage, one acre will produce about 100 pounds of dry grass for each percent of protein in the grass. That is, one acre of 40 percent protein grass would produce 4,000 pounds of dry grass in 50 to 75 days from the date of planting. Thus, one acre of 40 percent protein grass has a potential value of $3,416 as its yardstick value and would provide more food value than $15,000 worth of fruits and vegetables (at 1945-1972 prices). Ten million acres of 40 percent protein grass would be worth twice as much as all the human food now produced on 285 million acres. Such figures stagger the imagination.

The size of the job ahead becomes apparent, however, when we realize that probably less than 10,000 acres in its present condition will produce 40 percent protein grass. Fortunately, that acreage is scattered all over the United States and will be found in small patches where an old manure pile or feedlot has been. Not all 100-bushel corn land will grow 40 percent protein grass, but good land will produce 50 bushels of corn or oats long after it will grow 20 percent protein grass. That is why we have not realized how far soil depletion has gone in the United States. Probably 20 percent of the cropland can be made to grow 40 percent protein grass, but only 3 percent is needed for a total solution to our nutritional problem, if it can all be preserved at the peak of its quality. *That 3 percent, however, must be the best grassland in America.* It behooves every farmer in America to see if he has any of that kind of land.

Here is the way to find out: Select the best *square rod* of land on your farm and put a ton of manure on it. Sheep or poultry manure is the best. Mix the manure thoroughly with the top 6 inches of soil, and when it is well rotted, plant any of the cereal grasses or

Sudan grass in season. In less than three weeks you will have to water the spot if it does not rain.

On the day the seed culms form their first joints, pull up about 100 culms, leaving as much soil attached to the roots as possible. Dry this handful of grass in the kitchen oven, being careful not to scorch any of the leaves, and then send it, roots and all, to the nearest laboratory equipped to make a complete soil and plant analysis. If the sample tests 40 percent protein, you can rejoice, because at least 10 percent of your land can be made to grow 40 percent grass at a nominal cost. If it does not test 40 percent, the right kind of laboratory can give you some valuable clues about what to do next.

The soil can't be too rich for grass, but soil physics is just as important as soil fertility. The 4.5 percent protein oats mentioned above were growing on a heavy clay that swells about 20 percent when wet, thus shutting out all air from the roots. Nothing can be done with that kind of soil. The ideal grass soil is a sandy loam with lots of nitrogen-rich organic matter, which will hold about 30 percent water and still have 20 percent of air space extending down to a depth of about 4 feet if the land is level in order to assure proper drainage in wet weather.

Remember that it does little good to grow 40 percent protein grass if it cannot be cut or grazed within a few days of the jointing stage. Wet fields are the greatest hazard to harvesting at the proper time. On the other hand, lack of moisture at the right time is the greatest hazard to the growth of good grass. Since so small a part of the total cropland is needed for 40 percent protein grass, production farmers in humid climates should give a lot of attention to both drainage and irrigation. It would be folly to build up land to produce 40 percent protein grass and then lose the crop because of too much or too little moisture. A possible solution, at least on some types of soil, is tilling for sub-irrigation in dry weather. The method can be tested on a few rods at small cost.

Many feeding tests with grass of known quality indicate that very few grazing animals get over 5 percent of 20 percent grass in their rations, even in the spring. To get the full value of good grass, it must be cut and preserved at the peak of its quality. There are many unsolved problems in the harvesting and dehydrating and storing of unjointed grass, but the biggest problem is to find 40 percent protein grass to harvest.

All the grass dehydrators are located in the most fertile valleys of America, yet the average protein content of dehydrated grass for the past ten years has been between 20 and 25 percent. Even 35 percent protein grass is rare. The hopeful thing is that every year there are small patches of 40 percent protein grass surrounded by a sea of 20 to 30 percent protein grass. These little patches in almost any wheatfield where the grass is twice as green and twice as tall are reminders of what the whole field *could* be.

The sole purpose of this report is to awaken the American farmer to the possibilities of growing good grass for dehydration. Grass for dehydration must be produced within 10 miles of a dehydration plant, otherwise the cost of hauling the fresh grass becomes prohibitive. A radius of five miles, however, includes 64,000 acres, and the ideal location for a dehydration plant is in an area where 20 percent of the land will grow 30 to 40 percent protein grass in fields of 20 acres or more. There are many such areas which can be developed but it would take a lot of community cooperation to make 20 percent of the land grow 30 to 40 percent protein grass.

The Grass Enigma

The best possible explanation of the difference between grass and legumes came out of Finland during the World War II era. The work of Virtanen and associates proved that grass and probably all other legumes, in terms of carotene per plant per gram of dry material, reached their peak at the same time. The highest quality, the research revealed, could be harvested at the bloom stage.

There is a different growth pattern with the cereal grasses. There is a rapid increase in carotene at the 18th day, after which it decreases to half of its peak value on the 25th day. The plant's carotene continues to rise until the 46th day, the bloom stage, at which time it is almost withered. With 35 percent protein at the jointing stage, a bare 10 to 15 percent protein is reatined at the bloom stage. If wheat was planted three times thicker and grazed before jointing, it could be grazed at least three times, with three times the quality.

The course for vitamin A and D is much the same. Sulfur decreases some 66 percent between jointy and blooming. Ascorbic acid and glutamine occur together in jointy grass.

High protein grass harvested at the first jointing stage may be it for the pasture manager who deals mainly with perennials, but it drives home the point that fresh grass, not hay, not legumes, is the key to bovine health, a safe human meat protein supply, and fresh raw milk no longer dependent on the crutch of homogenization, pasteurization or irradiation. Grass confers on the grazing animal hormone and enzyme systems that cancel out pathogens not only in the milk, but in the manure distributed throughout the pastures.

Since the bovine is an herbivore, not a ruminant hog — as rancher Jim Lents identifies feedlot beef — reference to tests with smaller animals that are herbivores seem in order. The guinea pig is an herbivore.

Only in the last half century has it been determined that even one pound of jointed grass contains more vitamins than a ton of the same grass harvested as hay.

One early experiment consisted of feeding test animals on completely synthetic diets much as was the case with World War II prisoners under the management of I.G. Farben. Guinea pigs were fed rations that included minerals, proteins, carbohydrates and 13 measurable vitamins. The addition of jointed grass to their diet recaptured health. Also it was learned that 5 percent replacement addition of fresh grass to the ration made possible total elimination of any vitamin. The experimenters relied on high protein cereal grass, not pasture species.

The Cattle Connection

In evaluating cereal grasses fed at the first joint stage versus alfalfa, mature hay and other foods, grain included, a sample run of steers was pastured on the high protein cereal grass. It became evident that one acre of the most ideal grass put 500 pounds of gain on medium-weight steers in 30 days. The oats in question had been planted March 1. Grazing was started on March 20, and terminated April 20. The steers could not keep up with the growth, and most of the feed had jointed. The steers could not have eaten more than 2,500 pounds of dry matter. This means it took only five pounds of dry matter to deliver one pound of grain. Those steers topped the market as grain-fed beef even though none had consumed even one pound of grain.

Good grass was responsible for the best of conditions. Kansas City steaks were unsurpassed in flavor before feed lots became "theory period" state of the arts. Kansas City acres grazed more cereal grasses in spring and fall than any other area in the world. Fifty or 60 years ago, this was common knowledge.

An errant assumption that bedevils grazing suggests that anything green in the pasture forage will serve better than grain. As a test animal, the guinea pig answers questions no amount of chemistry can even approximate. Feeding tests dispose of what men propose. On a diet of head lettuce, death claims this test animal. Cabbage and carrots may keep it alive, but only at half the normal rate of growth. Yet this little animal, much like the bovine, thrives on grass.

All leafy greens are rich with vitamins associated with plant synthesis.

Early in the publishing life of *Acres U.S.A.*, one of my writers cited a USDA bulletin to the effect that when alfalfa is substituted for grass pasture, a sharp drop in milk production followed. The same writer called attention to the beneficial effects in the liver resulting from good pasture grass.

Citing USDA Bulletin 644, the term "green" was defined and the case made for attention to grass as absolutely the most valuable green on planet Earth. Greens are usually cooked for food whereas salad plants are eaten raw. Cooking annihilates 20-80 percent of the vitamins, perhaps all of the enzymes.

Ranchers still reach for the mineral box when they should be fertilizing pastures in order to enlarge the miracle they see each spring when animals suddenly take on a veritable glow as their coats become sleek. This miracle is achieved on jointed grass.

Grass delivers vibrant health, meaning the presence of still undiscovered food factors.

Grass/GHRA

The very term *grass* was derived from an old Aryan root, *ghra*, meaning "to grow." *Grain*, *green* and *grow* are kissing cousins, so to speak, all derived from the Latin *gramen*, "grass." Dictionaries usually miss the mark when defining something as complicated as grass. They tell of blades and leaves eaten by horses, cattle, sheep.

"All flesh is grass, and all the glory of men as the flower of grass," reads Peter 1:24. The grass to which the grazier delights the health and production care of his animals has a name, *Gramineae.*

All grasses have stems with solid joints plus two ranked leaves, one at each joint. Leaves have two parts—a sheath that fits around the stem like a tube that has been split, and a blade. Even seed heads have a character all their own. Flowers exist on tiny branchlets, sharing a crowded residence, always paired like the leaves.

Most grasses flower each year. There are exceptions. Some perennials are spread with the aid of underground specialized stems, rhizomes and rootstocks, and fail to flower regularly. Grasses are specialists at simplification.

Obeying the biblical injunction to increase and multiply is the name of the mandate in nature. Hidden deep in the ovary of the mother flower or between the scales of a seed cone is the ovule. This contains an embryo sac and an egg. The egg must be fertilized by a sperm cell from a pollen tube before it can start to develop into an embryo, and thus perpetuate the parent's life.

Much like any infant, the embryo needs a special store of food, a formula on which to live after it has become separated from the mother plant. That's why every seed has its ration of carbohydrates, proteins, fats and minerals. Just what the package contains is programmed by the computers of nature according to the kind of seed. Corn, as might be expected, is heavy in starch. Flax and sunflowers specialize in oils and fats. Peas and beans like protein. Seeds like to hide out their food stores in a diversity of places. Some stash the reserves inside seed leaves. Some place the goodies away in tissues developed from the embryo sac, in the endosperm, or in tissues developed from cells of the ovule that surrounded the embryo sac.

Nature has even programmed a distribution system for seeds. Some travel on the wings of the wind, fitted out with a feathery pappus that serves as both a sail and a parachute. Stick tights attach themselves to animals and hitch a free ride. There are the tumblers, and the passengers in alimentary tracts, which are impervious to digestive fluids and gizzard grinding. All seeds need proper temperature vacillation and environment or they won't grow.

Each seed a farmer deals with comes with a built-in computer. Nature's programming tells the plant when to sleep, when to wake up, how to translocate nutrients, and how far to go in the food-production business. As long as enzyme systems work properly, the seed does what it is supposed to do. It takes a trace element keys to activate these enzymes. If the farmer does anything to upset this fine-tuned computer system, strange abnormalities start appearing in plant, animal and man.

Even so, seed-producing plants — or spermatophytes — are merely an end product in a long chain of development. Affected by the evolutionary process were not only the reproductive, but also the vegetative structures of plant life.

Wheat Grasses

Wheat grasses are not the same as wheat grains. The former are the range grasses that tax students with Latinate names. These are native grasses that endure after others fade away. They are perennials with or without running roots. The scientific name says it all, *Agropyron*, from *agros* (field) and *pyros* (wheat). The northern Great Plains is wheatgrass country, as is the intermountain region.

There are genes and species too numerous to discuss in anything less than encyclopedic coverage. Certain names will surface here and there, but for our purpose the anatomy of grass is best defined by the benefits grazing confers on the ruminant animal.

Withal, poverty grass is impoverished because the soil is exhausted, usually because of bad treatment. Poor grasses, much like nutrition-poor people, endure, but they do not produce champions in either realm.

Keeping these few points in mind, we will now move on for visits with some of the top grass afficionados on the rural scene today.

Real Beef

Beef heifers grazing small pastures of Kentucky or Missouri-96 tall fescue varieties generally regained about 50 percent more weight than heifers grazing wider and more common varieties such as Kentucky bluegrass. Spring, summer and fall tests yielded this information after pressures in the pastures were kept as uniform as possible. Herbage was kept at about 25 percent body weight, 35 to 68 days. "We were testing the quality of the grass rather than its yield," noted agronomist Arthur Matches.

3

Dung Beetle Heaven

There is a ranch in Oklahoma operated by Walt Davis that might well be considered dung beetle heaven. Within hours after a cow patty has hit the turf, dung beetles on that ranch feed on cow platters, burying the rich, ripe materials in underground burrows, usually forming up a ball or rolling away a part of the deposit for interment. Few ranchers are cognizant of what is going on — in fact, it really doesn't happen too often nowadays because industrial agriculture has managed to annihilate populations of beetles even a well-paid workforce can't replace.

"Millions of hectares of pastures in the United States are lost to grazing each year because of dung accumulation and contamination." That statement anoints a report, *The Potential Value of Dung Beetles in Pasture Ecosystems*, published by Truman Fincher, then of the Veterinary Toxicology Research Laboratory, now closed down as an economy measure by the administration that hopes to colonize the moon and send lunar explorers to Mars. Fincher's term hectare need not detain us. Multiply hectares times 2.47104 to convert the mandated science of formal papers to usable-form English.

For the past 55 to 60 years, the population of dung beetles in pastures has all but vanished where toxic sprays have been used for weed control. As it stands today, surviving dung beetles in most pastures cannot remove the millions of tons of livestock feces deposited daily in paddock pastures or on open range.

For over three decades it was Fincher's job to select, propagate, colonize and in general nurture the objective of putting the unpaid insect workers back on the job. The program involved capture of native and foreign dung beetles for the purpose of salvaging thousand of acres otherwise unsuited for grazing until the slow rhythms of water and seasons melted down the contaminants, usually losing the economic value of the manure.

Three Species

By the time Fincher's laboratory was closed down, three species had been released in Hawaii, and Texas acres had been populated with results so successful they cried out for a national program. Releases on the drawing board would have created a massive complex for every major grazing area in the country. The objective was rapid removal of livestock dung from grassland surfaces, saving tons of nitrogen for agronomic use in the process. Fincher's abstract even computed the reduction of helminth parasites, pest fly populations and other aboveground competitors for life sustaining properties of fresh bovine manure.

In any dung beetle-rich pasture, one can observe these cow pie engineers feeding on deposits of cow droppings, always burying parts of the material into underground burrows, forming a hill in the process. The dung beetle rolling a ball made it into classical culture in Egypt, where the scarab achieved sacred status.

"These insects are playing a role in a lengthy food chain that began with the assimilation of energy from the sun acting on the green leaves of plants used as food by grazing animals" became the Truman Fincher summary whenever he was called upon to defend and explain.

Dung beetles feed on the fluids of fresh dung. They use whole dung to feed their young.

There are three general groups. These are classified according to food manipulation habits. The classical name for the Egyptian Scarab is tumble bug. You can see them pounce on fresh camel dung, seemingly arriving out of nowhere. The tumble bug rolls manure balls as if they were marbles. Beetles either feed on the fecal material or they deposit their eggs in them.

The second group can be characterized as a dung burying beetle. They take the manure to underground chambers at the end of a burrow where a ball is constructed to house an egg. This much accomplished, the ball is plastered over with clay or packed away for feeding or egg incubation. Topside, the manure pie disappears in a day or more, sometimes within hours.

Truman Fincher made it his life's work to study dung beetles, to propagate them, and to work with ranchers in colonizing them across the nation.

The dung feeders are of a different stripe. Simply stated, they feed on dung. They complete their life cycle in the deposit itself.

What follows is an abstract in depth of the life, work and findings of and by Truman Fincher.

Names A'Plenty

Dung beetles have names — as families, genus and species. The family name is *Scarabaeidae*, and there are some 30,000 species. North America has almost 1,500 species. All have been classified, dissected, analyzed and showered with wonderment, the biggest wonder of all being the indifference ranchers and scientists have accorded this most valued creature. It would take an encyclopedia the size of an unabridged dictionary to get into the bug classification game. Suffice it to note that Fincher finally settled on a few species for repopulation of North American pastures, first noting that there were species that would roll a ball of manure as big as a tennis ball. In his classic paper on beetles, Fincher tells of manure balls, coated with clay, that were found by archeologists during digs and were believed to be ancient cannon balls.

The value of dung beetles to pastures and the ecosystem was first acknowledged by George Bornemissza, a Hungarian. Wrote Fincher, "The viability of every pasture eco-system is based on the normal function of its nutrient cycle." The grazing animal is one component. Its role in keeping the system operative is acknowledged by all who think first of ecology before they turn to salvage via chemistry. In effect, the bovine animal is its own manager of its environment if allowed to function with an ample supply of air, water and forage presented as living room. Mismanagement brings on repercussions.

The concept of introducing dung beetles into pastures from other countries was advanced by Bornemissza in Australia, a nation that ranks head and shoulders above the United States in the introduction of the beetles. Most were harvested in Africa, a continent rich in animal species and beetles to handle the large inventory of manure droppings. At issue, then and now, is pasture wastage or fouling by dung.

Bornemissza wrote as follows: "The Australian grassland ecosystems were profoundly disturbed by the arrival of domestic stock. Prior to that, the nutrient cycle is thought to have functioned smoothly and that the dung of the principle herbivores, the marsupials, was relatively unimportant and probably never accumulated in polluting quantities. The portion of the marsupial dung was buried by the native dung beetles — this burial sped up decomposition and returned essential nutrients to the soil. The introduction and rapid increase of the number of domestic grazing animals progressively upset this primitive cycling. As the droppings of horses and cattle accumulated, the pollution problem arose in many areas — to the extensive fouling of valuable pasture land."

When the buffalo passed over an area, or deer and elk gifted their droppings — as did birds and smaller animals — native dung beetles and other coprophagic organisms easily handled their dung. The beetles that obeyed the biblical injunction to increase and multiply easily blended their function to the demand of nature. They removed the droppings from the soil's surface with ease and efficiency.

Interdicting Forces

Over the past 50 years, the near-universal use of salt fertilizers and toxic rescue chemistry has turned the "American Serengeti" into a dung beetle desert. New grasses and greater yields have increased carrying capacities and therefore a new inventory of dung awaiting disposal. The dung is ready for transport underground the minute it hits the turf. Unfortunately, that is not what often happens. Cows refuse to graze spoiled grass, and spoilage takes place over a wider area than the cow platter, an area often as big as a baseball home plate. With beetle population all but gone, cow pads often stay put for months.

Speaking before a San Antonio assembly of *Acres U.S.A.* readers, Truman Fincher mused about the apparent and real decline of the beetle population in Georgia.

"Many older farmers in Georgia have stated to me that there are not as many tumble bugs today as there were when they were growing up. However studies of the dung beetle fauna on the coastal estuary of Georgia reveal that thousands of beetles can be captured within a few days, including tumble bugs. I am amazed at the number of pairs of *Canthon pilularius* that I observed daily on Ossabaw Island at a single observation point. There were rolling balls of cattle dung in an open area of an acre. After 12 years of trapping beetles in Georgia, I have never seen this many tumble bugs working at the same time in one place."

Since no crops are produced on the Georgia islands, no use of farm chemicals has been the norm. Moreover, area livestock are not treated for parasites or pests. New forages, chemical insecticides, intensive grazing and some few modern husbandry practices all have conspired to depopulate dung beetles on the mainland.

Clearly, an overload of undigested manure on pastures reduces the available grass, and this translates into ever-faltering figures on the bottom line.

During his more than two decades as the nation's only dung beetle aficionado, Truman Fincher made countless calculations on acres recovered from grazing and economic losses, such as the nitrogen lost when dung isn't melted away, captured by beetles or removed. Rank growth around each cow platter costs as much grass as the area covered with dung. Horn flies and feces flies move in — rather they breed in fresh dung. Undisturbed, the cow pie becomes a residence for various stages of gastrointestinal parasites that torment and debilitate livestock. The nitrogen content in the cow dung goes up, not in smoke, but as an unseen miasma. Other nutrients, research has revealed, are taken up for months, even years, and not available to feed the organisms that fed the pests.

The Stocking Rate

Grazing animals at a stocking rate designed to consume the available forage that fertilization is capable of yielding courts the danger of poor animal performance. Some cattle producers have sought to make an end run around the surplus forage problem by making pellets, stacking hay and otherwise servicing the feedlot. In nature the world energy situation has endured a paradigm shift, and cattle raising that depends on expensive machinery and fossil-fuel energy consumption is no longer defensible, on paper or in fact. Thus the beetle comes back in focus and feces removal, a technology devoutly to be wished. Harvesting and pelletizing forage is too expensive and too unscientific nutritionally.

Sanitary Engineers

Nature's first sanitary engineers were the dung beetles. These dung rollers have biographies that start with the Pleistocene and roll forward through climate changes, volcanic upheavals, even strikes of meteorites from outer space. Likely, North America never had nor could have had the proliferation common to tropical Africa or South America. If snow and ice pushed mammals north from the south and south from the north, it's certain the beetles followed their source of food.

During his tenure at College Station, Texas, Truman Fincher made several counts, a few of which are of interest to North American graziers. Data available to the researcher revealed approximately 1,137 species of *Scarabaeinae* in the Western Hemisphere. Only 945 live in South America, 157 in Canada, and 197 in North America, Mexico included. Some overlap geographic boundaries of those counted for North America, 130 reside in Mexico, 87 in the United States. Of the 87, only 40 percent are in the high grazing areas of the country. Only three species range across the Southwest.

In all probability, the contradiction between pesticides in the pasture and a healthy crop of dung beetles suggest an oxymoron. But the possibility of manure being handled by dung beetles in a healthy pasture eco-system invited a national attention the farm industry never got. The token laboratory and its work became a casualty to a budget cut, as explained earlier.

Cattle on Farms

It is next to impossible to estimate the number of cattle on farms and pastures any one year: the number has to be in excess of 110 million calves, heifers, steers, bulls. It can be estimated that such a population would deposit at least one billion cow platters every day of a 365-day year. Many of these fecal deposits remain in view at least six months. Any way you count it, a guess would be ten cow patties each day, making it possible to compute the overload for a single pasture or farm. College Station research revealed that 82 to 88 percent of artificially deposited 2,500-gram cow pads were buried within a week during a month of peak dung beetle activity, *Onthophagus gazella* presiding. By way of comparison, 96 to 100 percent of feces deposited in March through June, with only native beetles present, remained on the pasture surface after one month. Of feces deposited during winter months, 75 to 95 percent remained in the pastures after nine months, 0 to 40 percent after 16 months.

The lesson is clear. Losses damage — if not stagger — those trying to reach the bottom line while still using toxicity. Otherwise, propagation and colonization of the right beetle species remains unfinished business for the Department of Agriculture.

The Damage

One of the researchers mentioned earlier concluded that even one cow pad damages an area one meter square, a little less than a square yard. On dry feed the damage is 0.8 meters square, 1.2 meters on silage. These numbers are the most conservative of many studies.

Hypothetical premises are just that, hypothetical. Just the same, give-and-take numbers should suffice for the points being entertained here. That national cow herd mentioned above — give or take — should cover 12,288 acres of pasture per day, or 4,485,295 acres per year.

Give or take tells us to subtract when grazing stops, and it tells us to make computations that differentiate between cows, steers, calves and bulls, defecations moving from six for a calf to ten for a cow. Multiply the grand total times 365, and the total of acres involved that are out of commission rises up to haunt and horrify.

Some 13 million cattle are always in feedlots and do not graze. Evacuating droppings in and around farms, in woods, etc., and assuming 75 percent of the dung is dropped on pasture, and half remains on the pasture surface for several months, then it appears

— says Truman Fincher — that 829,474 acres are continuously covered by cattle dung each year.

Undigested dung on grass is at least as damaging as noxious weeds. Cattle will not graze on rank growth, nor will they eat contaminated grass. Cattle refusal of contaminated grass is often 80 to 83 percent refusal, this when there is only 16 to 26 percent refusal of the same grass with dung being absent.

It does not take a rocket scientist to understand why animals recognize the offensive properties of dung. First, it imparts an undesirable taste to the forage. If an animal can satisfy its needs in any other way, it will keep a wide distance between itself and manure still drying in the sun.

Research workers have wide differences of opinion between each other over how wide the cow pad delivers contamination. As mentioned earlier, figures used by Fincher's College Station research always trended toward being conservative. Using the computation which assumes some 829,474 acres are covered by dung each year on a continuous basis, Fincher computed that 4,147,369 acres of grazing area are constantly off duty to grazing because of this contamination. Assuming 96,075,000 cattle other than dairy animals would effectively remove 4,147,369 acres of pasture from production each year.

Guesstimates are guesstimates, no more. Still, the cost has to be astronomical. There are calculations that translate such losses into beef on the hoof, but the point here is pasture loss, and the economic losses can be settled farm by farm with a small calculator. Prices and each passing year make calculations suspect each day. The dairy farmer can construct his own numbers and measure his own profit leakage, this assuming dairy cows are maintained chiefly on grass and not fed high energy rations, and escape partial or total confinement.

Nitrogen

At least 80 percent of the nitrogen contained in cow dung evaporates and is lost to ground cover. The literature is full of citations and footnotes by workers who have tackled the subject. This loss can best be mitigated by the presence of dung beetles. Herds stomping these nutrients into the soil before moving on help, but dung beetles when colonized represent nature's finest development.

One researcher duly noted that most of the nitrogen is present in a component of undigested protein which is denatured by bacteria. This is lost by utilization as ammonia. Buried by dung beetles, that same protein envelope has a beneficial effect on grass growth.

Using a factor of ten eliminations, 25 kilograms (55.11 pounds), the payback has to be measured in tons per day, whatever the national animal count that year might be. The total per year has to be in the neighborhood of 436 or 438 million tons per year. At 80 percent moisture, 2 percent nitrogen, on a dry-weight basis, a single year of Fincher's calculations meant the loss of nearly a million tons of nitrogen, more or less, representing the economic loss of N.

Failure to support the facilities needed to repopulate the nation's pastures with dung beetles reveals a shortsightedness in public policy that goes beyond criticism. The cost of replacing such a nitrogen deficit runs into the millions even when computed at yesterday's prices.

Control Agents

There's more. Dung beetles are a valid control for gastrointestinal parasites. Parasites, like the poor, are always with us, especially in warm climates. Parasites mean morbidity and veterinary costs, medicines included. Parasites bow to sanitation, and again this means they cave in when a good cleanup crew is on duty early. The near wipe-out of these natural janitors in the pasture due to changed technologies over the last half century assigns cause to effect. It may be that the range fed animal has less of a parasite problem than the resident of a tame pasture or the animal that runs low on feed in a rotational system as feedlot animals readily illustrate, crowding augments the transmission of parasites. Any concentration of host and parasite in a small area holds in escrow the danger of infestation. Control with pharmaceutical chemistry does not fall under the purview of this narrative. A new look at the efficacy of dung beetles does.

When eggs are passed out of the living host and then allowed to hatch in undisturbed dung, the parasite takeover is off and running. Moisture and temperature preserve infection. Migration of the infection to grass becomes a given. The grazing animal attempts to avoid the infestation, but the life cycle of the parasite completes itself unless something interferes with it.

This approximates the dung beetle story. Researcher Truman Fincher, now farmer Fincher, will nod his head in abject disbelief over the monument to the stupidity of man that cancellation of the dung beetle projects has become. Valued research tells of a nine-fold decrease in infections when pastures were given new residents, dung beetles. Losses invite the same calculation already expressed for nitrogen, grazing areas, etc. Mortality due to parasites is generally reckoned at 0.25 percent. Mere debilitation is not easily measured. All such arithmetic merely hints and staggers. It never makes the *Statistical Abstract of the United States.*

This much stated, how can it be that manure is a most valued product for the diversified farm, even for the pasture? A codicil of sorts may belong here, because the quality of manure affects grass, pasture and the paddock environment. Mark McAfee of Fresno, California, is a dairyman. He brings his milking parlor into the field and preserves the health of his animals with sound husbandry.

"The bottom line," noted McAfee, "is that once you stop the physiological stress, you stop the production of pathogens."

This comment seems self-evident, taking factory herds into consideration. Great inventories of antibiotics are used, as are hormones. These cattle populations are identified with poor manure management, no natural grass, and absolutely no way for the cow to really rest. Because of physiological destruction, factory cows slough off pathogens in manure constantly.

Catherine Berg of the University of California, Davis, is a veterinarian. Her tests revealed that grass-fed cows at Organic Pastures dropped manure that contained absolutely no salmonella — this finding from manure directly evacuated by cows. That is exactly opposite of results for the infected factory herds forever bedeviling Congress for bailout money.

McAfee added, "We took our milk to labs here in Fresno. They added 10 million counts of pathogenic bacteria to each liter of milk. They wouldn't live."

A few more answers emerge in the chapters that follow.

4

The Microbial Connection

The mystique of manure and ocean water will surface again as the pasture endures scrutiny. In the early 1970s, Fletcher Sims of Canyon, Texas, nudged commercial composting into existence. It didn't take long to discover the efficacy of well-digested compost and compost tea on faltering pastures. Even so, the mystique of manure — even with the lessons of Rudolf Steiner and Ehrenfried Pfeiffer in tow, even with an ocean water connection — posed as many questions as its answered. In an oblique way, grass proposes and nature's requirements do the disposing. In the mix holistic resource management (HRM) has quite a bit to say.

The findings of Rudolf Steiner are well known, and few seekers doubt that the livestock in the soil outnumbers everything that grows above the ground. Use of hyperactive materials for foliar sprays and soil amendments is common these days, with well-designed packages annihilating the claims of salt fertilizers through nothing less than performance.

Tracking back through the pages of history, I came across a pioneer, Arthur Franke of Hondo, Texas, and then the story unfolds in depth. Pastures and crop acres are the beneficiaries. No doubt there are sidebars to the story, but for the purpose of this chapter the years 1950-55 have to come into focus. By then James Francis Martin was well into an approach that has left its stamp on pasture management. What follows belongs in the "lest we forget" category.

Martin has been characterized as an inventor of the Thomas Edison stripe. *Texas Monthly* described him as a self-taught chemist, metallurgist and naturalist with a fourth grade education. He was a rational man, with a passion for discerning how things worked. That included plants, the soil and the foods plants and soils produced.

He discovered natural insect repellants for plants. He invented a pollutant-reduction muffler. He even figured out how to make synthetic opals.

What he discovered was barely accepted during his lifetime. He died in 1975. He had developed an odorless, colorless *living water* — his term — using the raw materials of seawater, cow manure and yeast. He used fermentation to transform his mix, capture all the trace minerals and stimulate microbes to obey the biblical injunction to increase and multiply. His end product could cleanse polluted water and soils. His living water could dine on sewage sludge and cause the desert to deliver abundant crops. Yet Jim Martin demonstrated and preached to an agriculture that wasn't listening.

Yet today several Texas companies, and several located in states as far-flung as Pennsylvania, Florida and Missouri are offering what some historians label "variations" of the central Martin idea. Some 30 years of reporting, seminars, conferences and one-on-one conversations have caused these products to be studied seriously. One variant, coupled with sound vibrators, is making significant inroads in the heavily populated areas of the Pacific rim, Hawaii and select farms in the state side U.S.A. The capability of increasing insect-free crop production has been noted and reported. This has left unanswered questions of "where from," etc. A capacity to reclaim salt-stressed soils has been noted, and there have been reports of crop thriving with salt water irrigation. Most important, studies under the auspices of the Department of Energy have confirmed that microorganisms treated with products akin to Martin's water can enhance pasture production.

Whatever the disclaimers, the general idea can be traced back to Martin. A few trade personalities alive today remember Martin as the "old man." Even his name has retreated into the archives.

Not so with Martin's findings. Entrepreneurs have adapted those findings to crop and pasture production, sewage treatment, and foliar nutrition. This is not to say that everyone has forgotten the old man. Probably the greatest student of the old man was Arthur Franke. In the twilight of his life, Franke recalled some of the old man's premises and a lot of his conclusions.

A reference to J.E. Lovelock's *Gaia: A New Look at Life on Earth* seems to be in order. This is the famous book that explains how Earth's life came into being, how the planet balances itself, and how the secret of life has unfurled. From the hands of this noted scientist came rhetoric so much in line with Martin's findings, both Lovelock and Martin have to be considered Gaia's children.

It all has to do with methane and ammonia and oxygen enveloping the planet. The old man believed he understood the process and he duplicated it. Martin patented his process and offered it to a world that needed to undo its damage to the Gaia called Mother Earth.

One of man's follies is revealed in the oil outflows that occur when tankers rupture, as was the case in Santa Barbara in 1969. Carl Oppenheimer of the University of Texas at Austin was appointed to the special commission dealing with the spill. He was hard on the hunt for microbes capable of cleaning up the oil spill. When President Nixon named him to the study panel, Oppenheimer had studied single-cell ocean-living organisms. The professor had been salvaging bugs from around the world, propagating them and assigning them the task of gobbling up spills. His problem — it took a month of Sundays to get some real work out of his bugs. While on a hunting trip, Oppenheimer met some Texas oilmen who knew the old man Martin, and with that connection the hunt for the old catalyst was off and running. The locus of attention fastened itself on Hondo, Texas, an agricultural hub of 6,000 people at the time. It turns out Jim Martin had moved to Hondo around 1950, passing away in 1975.

It seems a politico named Tom Kurykendall, a sometime partner to Martin, was a shareholder in a company named CLEW, which paid Martin a modest fee as they tried to market the product. CLEW was formed in 1971. A fermentation tank was set up to supply a creek in Memphis with water that had been fouled for 30 years by oil runoff. It took seven weeks to return the lagoon to health.

There was a waste lagoon near a packing plant at Clarksville, Tennessee, with a foot and a half of grease. The old man's formula dispatched the pollutant, but even eyewitnesses refused to believe what they saw.

Oppenheimer was introduced to Kurykendall, who asked the scientist to test the old man's product without asking how or why it worked. The professor did make a test. He found that Martin's invention made microbes reproduce at about 1,000 times the normal rate. Moreover, the progeny of those base organisms could survive in inhospitable terrain. The prospect of recovering petroleum in played out fields and wells was a speculation the schoolmen had entertained for years. ALSA came into being. It was dedicated to oil spills, chiefly the Alaska trouble.

This industrial application of living water seems to have further eclipsed the legacy of the old man, born December 13, 1894, in El Paso. At age 18, Jim Martin went to work at Southern Pacific Railroad. When he retired from the railroad, he found time to develop his living water. Patent number 2,908,113 was applied for in March 1956. It was awarded in 1959. Martin assigned it to Arizona Actite, Trelson, Arizona. As if to draw a lesson from Rudolf Steiner, Martin used filtered cow manure — manure of milking cows — with fresh seawater to generate blue-green algae. He further refined the product to create phytoplankton. Manure, seawater, yeast, dilution and harvest of enzymes became the end product.

The Trelson labs are now abandoned, even though surrounding soil is sharply out of place in this desert area. Agriculture wasn't ready to listen or even see what it looked at.

Why milk-cow manure? Cow manure contains methane, ammonia and unique bacteria. Lactating cows make an exceptional amount of calcium for milk and calf sustenance. The amount of bacteria found in milk is also found in a cow's stomach. These bugs help cows take calcium from grass. These bugs break apart organic molecules. Martin relied on these organisms to power his stew. It is now speculated that Martin in fact cultured an ancient form of bacteria called photosynthesizing prokaryote not unlike those living three billion years ago. The bugs involved needed hydrogen and photosynthesis, and they got it from the water by splitting the water molecule into its component parts, two parts hydrogen and one part oxygen, hydrogen being used for photosynthesis, oxygen a waste product. Oxygen achieved critical mass exactly as described in *Gaia* by Lovelock.

Martin's biography has yet to be written. For now it is enough to cite his notebooks, his trips to several lagoons, to pasture acres, his last stop being Hondo, Texas. His notes of tests that were 100 percent perfect, this at a time when USDA blessed toxic technology and gave it the imprimatur of high science.

Martin took one test to a small farm near Clint, Texas — Thurston's, near El Paso's lower valley. This soil could not be plowed, it was that hard. Martin treated the soil, and it softened. Martin made cakes out of spent manure and placed them in the irrigation flow feeding cotton crops. This manure tea worked so well that prime cotton plants reached above Thurston's head. Results on corn were equally astonishing.

Marketing such a development faltered, not so much because of unreal expectations, but instead because of the period theory/instruction that held land grant colleges, and therefore farmers, in thrall.

Businessmen exhibiting greed beyond the dreams of avarice vigorously cheated Martin. The Arizona company dissolved. Government agencies, their professionals defending their worthless bibliographies, simply cut the legs out from under meaningful ventures.

In the early 1970s, a new line of firms came onto the scene. By then Medina of Hondo, Texas, had installed and proved the Martin premises in the agricultural arena. There was Alpha Environmental, Austin, Appropriate Technology Ltd., Dallas, Spray and Grow, Houston, APR Products in Dickenson, and Bio-Plus in Hawkins. The idea was spun off to Jim McHale and Agro-Spon, then Nitromax in Dillsburg, Pennsylvania, finally to still others too numerous to mention. C.J. Fenzau and Don Schriefer ordered the fermentation tanks into operation producing Lactobase at Storm Lake, Iowa.

All the products will claim their special attributes, but few will deny the Texas contribution to the general idea — fertility without salts or a requirement for rescue chemistry. Valid technologies have been used to make foliar feeds and liquid fertilizers for crops and pastures since 1951. These developments went on virtually without the mainstream agricultural community knowing anything about them. Sylvan H. Wittwer, under whose leadership

Michigan State University conducted investigations on foliar fertilization of plants, summarized correctly: "There is probably no area in agricultural crop production of more current interest — and more contradictory data, claims and opinions — or where the farmer in practice has moved so far ahead of scientific research." Even today the average land grant college does not have many scientists who have a working knowledge of using liquid fertilizers in growing crops. There will be those who object to this statement, citing Hanway's Iowa State University research, which came some 12 years after Wittwer's, but this hardly qualified because it failed as a result of using the wrong type of ingredients.

Credentialed and uncredentialed outsiders standing in the wings knew that Hanway's foliar N, P and K efforts would fail, because the compound being used failed to answer plant requirements and would burn crops — it also cost too much. It involved the application of gross amounts of nitrogen, phosphorus and potassium to soybeans in three applications, using amounts required by holy writ of soil fertilization. As Albert Howard would have pointed out, fertilization of such crops entailed a knowledge of many other things, such as the plants' requirements at each point in the growth cycle.

Working with nursery crops, flowers for market, strawberries, apples, pears, cherries and even grass, agronomists started answering these growth cycle requirements with seaweed extracts many years ago, all with successes that were spectacular, erratic, and rejected. There were things that could be done with the leaf that proved next to impossible when working with complex soils.

Some few farmers knew this, but they didn't have a handle on the "why." In the early 1950s Wittwer was working under arrangements with the Division of Biology and Medicine, Atomic Energy Commission. Experiments entailed the use of radioactive isotopes in assessing the efficiency of foliar-applied nutrients compared to soil applications of those same nutrients. It was found that the efficiency of the foliar fertilizers was from 100 to 900 percent greater than the dry-applied fertilizer materials. The results became a matter of record under Contract AT (11-1)-W,1969, and soon touched base in select circles as an audio film entitled *The Non-Root Feeding of Plants.*

Natural Oxygen Products of El Paso — and perhaps dozens of others — disclaim any Martin connection, and it may be that a similar idea surfaced quite independent of the old man's track.

Hondo journals covering the 1950s era tell how the old man and a sidekick named Floyd Leland met Arthur Franke at Hondo, Texas. In 1960, the three incorporated Medina Agriculture Products (currently run by Stuart Franke, a nephew of the founder). The record isn't all that clear. The godfather of the operation was found by a *Texas Monthly* investigator. Medina put up the appropriate buildings and vats. The product that emerged was rated as a good soil activator, one that fanned the microbial fires to liven up the soil. Franke cited a proprietary formula, a staple industry statement. The Franke-Martin conversations in a local café were not recorded, but an abstract recalled years later revealed anew the findings of Albrecht, Hopkins, Bear and the Friends of the Land at Louis Bromfield's farm. It was the life in the soil that governed and presided over the condition and productivity of crop land and pastures.

Whether Martin developed a formula or Franke put two and two together based on giants who offered their studies to succeeding generations may never be documented. Franke in fact ponied up the money to start the venture.

The test crops proved phenomenal. Grain weighed in much heavier than normal. It paced the sale of excellence in Medina County.

On the basis of field trials, Franke, Leland and Martin headed the grand experiment, a process that could have turned the disaster of agriculture around.

Hurricane Carla had just totaled the Texas coast, September 1961. The grass called rice between Houston and Beaumont was in ruins; heavy doses of salt water having delivered mischief. The 1962 rice crop promised to exacerbate the reign of Carla's ruin.

Floyd Leland took several 55 gallon drums of Medina's product to the disaster area. He hired a plane to fly on a treatment for 80 acres. To the surprise of almost everyone, the rice field germinated. This invited other applications to dead areas. This injection of life caused fields to flower and crops to grow. The Medina application proved a whopping success across the board.

Carla proved to be a shot in the arm with a healthy tally resulting on the bottom line. The usual rubrics of marketing were invoked. A sample went out to Dewey Compton, a call-in garden show host. He gave his report on the Houston-based KTRH radio station. By the mid 1970s, the firm had 2 million in gross sales.

That's when the agency arm of commercial agriculture struck. The case reports harvested from growers across the country were dismissed as fiction. Texas A&M issued and distributed a report entitled, "Effect of Soil Activators on Crop Yields in Activation of Soil Microorganisms in the Southern United States." It is not known whether the same authors who wrote the paper also wrote the title. It was a hatchet job. It said that neither Medina nor any of the Martin-influenced products increased activity in the soil via causing microorganisms to increase. Medina was ridiculed as Hondo Holy Water. Sales plummeted. All studies that contradicted A&M were swept aside, even when the researcher had the credentials and standing required by academia.

The tag *snake oil* was laid on everything outside the NPK paradigm.

Medina barely survived a 75 percent decline in sales. It took years to recover ground lost because of a spiteful publication out of a land grant school.

Slowly, younger professors are rejecting the worthless bibliographies of their forbears. One at a time, and in groups, reports confirm the efficacy of the Martin-inspired products. Medina has racked up impressive research reports. A young professor has ratified the efficacy of Spray and Grow, a Martin-inspired product sent into trade channels by Fred Steffel of San Antonio. Retired schoolmen have been drawn into the fray, some of them previous employees of Ortho and similar firms.

A signal observation is in order. No one researches N, P and K any longer. Its so-called benefits and faults are too well known to require exposition. The rush is on to document the road back. The yields are still being improved by double-digit studies, but more important, less land is going down the Red River or other streams wherever Martin-inspired technology is invoked.

Martin was no businessman. His passion was to transport his ideas into the minds of all who would listen. He inoculated companies right and left with his ideas, for which reason there is much confusion over whether the old man was a prophet or a messiah.

Names like Medina, Agri-Spon and Nitromax are only a few of the down-line refinements of the thinking an old man accounted for in the first place.

Jim McHale, a speaker at the first Acres U.S.A. Conference and former Pennsylvania Commissioner of Agriculture, Commonwealth of Pennsylvania, was one of the first to accept the premises of eco-agriculture. He formed J & J Agri-Products in the 1970s, taking on Agri-Spon as a flagship product. Later he converted a Martin-style formula to Nitromax, a brand also sold out of Pennsylvania.

USDA's belated recognition of the microbial workforce came styled as the *Soil Biology Primer*, written largely by Elaine Ingham and associates, saying in science what old man Martin, Rudolf Steiner, and organiculture icons knew all along.

5

The Albrecht System

The Albrecht system works fine on corn and all the crops President Eisenhower called political, "but it *really* works on pastures." The quoted part of the preceding sentence comes from farmers who have abandoned the folklore suggestion that what won't produce corn, soybeans, milo and other basic storable commodities is fit to be left to pasture and left alone. "Not so," says Neal Kinsey, Albrecht's best-known practitioner and one of America's foremost instructors on pasture management. The lessons William A. Albrecht enshrined in literature and on the farm are as valued in the tropics as in Canada, on pastures as in fields of sugarcane.

Grass, after all, depends as much on soil as it does on sunshine and water, with air and CO_2 thrown in for good measure.

In a manner of speaking, the soils of Oklahoma, Kansas, New Mexico, Vermont, California and every other state define the hopes and the problems of the grazer.

The Anatomy of Soils

Kinsey knows soils the way an old salt knows the seas. It is a soil sample he wants to see first when arriving on the scene. A soil sample suggests a test, and it asks for correct definitions of the nomenclature used to listen to what the soil is saying. Thus the term *base exchange capacity*, sometimes called *cation exchange capacity*, CEC.

"Total exchange capacity," instructs Kinsey, "means measuring the size of your soil in terms of its ability to hold fertilizer." Accordingly, some soils have the capacity of a teacup, some have the holding capacity of a big bucket. Kinsey has to deflate preconceived notions. Listen to a standard presentation, "This soil from San Antonio has a greater exchange capacity than the best soil in Iowa," he says, screening samples from verdant fields. All the above in terms of exchange capacity of the potential of the Texas soil to catch and hold nutrients. The term "potential" demands "what is" because it often takes a wealthy person to answer the demands of holding capacity if perfect balance is to be achieved.

Kinsey tells of a rancher near Durango, Colorado, who ordered a realtor to find the worst acres in the area. He was a biologist. Having bought the land on the cheap, he wanted to reclaim the value the soil was capable of with the purpose in mind of growing grass. There was grass under irrigation. Once the samples were analyzed, Kinsey was forced to tell his client, "I have bad news." The bad news was the expense unbalanced land usually requires in order to achieve balanced, nutritious grass.

The client knew that remedies would be expensive, that conventional materials might be indicated, but he really hoped to do the job with natural materials. Fed into the computer and carried down to the bottom line, natural materials would be even more expensive because of sourcing difficulties.

The ranch surveyed out as 400 acres, about 135 under irrigation. The cost for fertility balance was to be $305 an acre, this for one year. Later, Kinsey had the opportunity to examine the records. The first year of intervention with nutrients based on sample readouts increased grass production five fold. The grower had computed that the sale of hay that year and the two years to follow could pay for the whole program. A buyer from Arizona pronounced the timothy hay off that farm the best in the Southwest.

The worst drought in 110 years followed. Now touch-up work was indicated. Production for the drought year came within 10 percent of the normal year, this with one supplemental irrigation. After that drought year, legumes started coming back even though no seeds were added to the turf.

The seeds always seem to be there, held in escrow by nature pending arrival of the right conditions. When fertility is made right, nature seems to give permission for life.

A certain sign that pastures are rich in nutrients is the arrival of wildlife, elk, deer — especially elk, which usually stay high in their mountain strongholds. Elk, deer and cows, Albrecht used to say, are better nutritionists than all the Ph.D.s in North America.

The point in reciting this case report is that the bucket was large — to continue the metaphor — and most of the nutrients pre-treatment were missing.

This holding capacity is seated in the smallest organic particles, humus, and the smallest clay particles, called "colloids." Take the mineral parts of the soil, sand, silt and clay, and make size comparisons, then the sand is a 747 jet, the silt is the size of a bald eagle, and the clay is the size of a humming bird. Soil colloidals are so small they cannot be seen with the unaided eye.

This capacity to hold nutrients is expressed as parts per million (ppm), or "pounds" in a typical soil report. Calcium, potassium, sodium and magnesium are thus converted from ions into pounds per acre or parts per million.

pH

The second factor that ought to trap the mind's eye is pH. Kinsey goes there second, after reading the cation exchange capacity, because the cations in an Albrechtian equilibrium automatically create a subtle pH, usually only slightly acidic. Most people test for a pH first, and if it is 8 or thereabouts, they think they have died and gone to heaven. Their error is to think that a high pH excuses concern about calcium, magnesium, sodium and potassium. Too often a pH of 8 is taken to mean there is too much sodium. To make his point, Kinsey often goes to a San Antonio soil sample. "If this man doesn't find salt, he's in trouble." Too little sodium is as troublesome as too much, and pasture grass expresses that imbalance with devastating finality.

A pH of 8 does not mean the absence of a sodium problem. Quite the contrary, a complex cross index of information packages provides the answer if the farmer reads the numbers on that soil report.

Organic Matter

Of maximum importance is the percentage of organic matter, meaning the organic matter complex of the soil called humus. It is not a compilation or measurement of leaves, clippings, roots and other materials subject to decay. Humus attracts and builds nutrients, a task other organic matter can perform after suitable decay has taken place. A good organic matter readout becomes holy writ when it reaches 3.9 percent because at that level the natural nitrogen cycle is operative at 98 pounds. Yet that nitrogen figure is not a measured amount of nitrogen in the soil. Compost, hard N_2 or manure a week or a month ago, it will not change that number. The soil audit shows how much nitrogen is going to be released from the humus during the pasture's growing season. If the crop is soybeans or clover, it is no different.

Consultants around the country tell us that farmers who religiously check their bean and corn growth are reluctant to monitor grass ground. This probably goes back to the idea that "this ground won't make corn, but it's good enough for pasture." Actually it is not, for which reason Florida cattle can graze cosmetically beautiful grass and be in a starving mode.

The foundation for herd health is pasture, not fabricated feeds or the mineral box. Alone among the hundreds of crops American farms grow, grass is the one that can pick up the most nutrients, if they are there!

Admittedly, cattle and horses and goats distribute their manure, a fertility blessing, but eventually the calcium will leave as pH goes down. As acidity takes over, the nutrient base nutrients are lost. Calcium, Albrecht said, is not only the prince of nutrients, it is the king. It is opaque and solid, and the other major nutrients — N, C, H, O, for instance — are gases. Failure of calcium to occupy 65 to 70 percent of those colloid positions interdicts the growth of nutritious grass and forage.

Neal Kinsey baffles farmers when he discusses acres he hasn't even seen. He has a calculation based on the calcium and magnesium revealed by a test. This information plus the nitrogen contained in the humus — subtract that pound for pound along with the nitrogen produced by the legume, and calculate the bushels of corn the going system will make.

In relating their pasture stories, farmers almost always drift back to row-crop experiences. Quite frankly, few farmers keep really good pasture records. Some ranchers and small cow-calf operators do.

Humus determines the stable nitrogen source. A 4 or 5 percent humus supply puts 90 to 100 pounds of released nitrogen on the other season side of the equals sign. That is why it becomes impossible to have an honest organic acre under the Organic Standards Act when humus hovers around and under 2 percent.

Grass takes up surface sulfur expressed as elemental sulfur in ppm, parts per million. This ppm equation is state of the art. You can take ppm and multiply by two to get pounds per acre. Thus a readout of, say, 6 equals 12 pounds per acre. The usual shortfall can be discerned when it is realized that at least 40 pounds per acre are nature's requirement for production of the best grass. If the sulfur number is 20 ppm, then obviously no sulfur is required, whatever the source.

Moving on, using the Albrecht system, the phosphate load comes into view. Phosphate is an anion. Punch a soil probe into most pastures and the readout index often reads a phosphate shortfall. The deficiency often runs into hundreds of pounds per acre. Without a knowledge of tests and circumstances, knee-jerk reaction can exacerbate the problem and elude the solution. Should the pH be well on the alkaline side of neutral, the said soluble phosphate test fails in producing the right answer. Anytime pH is above 7.5, a second phosphate test has to be invoked. This is the Olsen test. It is always expressed as pounds per acre in parenthesis. These tests get a bit esoteric. A water-soluble phosphate test, a P-1 test, is always a top line item without parenthesis!

The complications involved here might suggest the employment of a Philadelphia lawyer. Not so. But a good consultant can help. Once base saturation for calcium reads 74 percent, the P-1 test drops by 90 percent. For this reason consultants such as Neal Kinsey, whose insight on crop and pasture management is quoted here, a readout of 120 is considered very good for phosphate levels; 125 is considered excellent.

Most of the research available to farmers via Extension services was done with triple-super phosphates, or simply hard-rock phosphate treated with sulfuric acid and then with phosphoric acid to process off the sulfur and calcium, now 0-46-0, triple-super. This is the cheapest and the one most likely to be offered by the supplier.

There are dangers when messing with chemistry. One farmer we know used sulfuric acid to bring down the calcium level. Whatever the apparent result, such business would deliver mercury to the plants and finally to the animal and human population. The change to microbial activity can hardly be calculated.

Four to eight weeks after application, triple-super reverts to hard-rock phosphate, or tri-calcium phosphate. If the pasture is not correctly positioned to process 0-46-0 the plant won't pick it up. Slides from as far back as 1950 by TVA have made these findings a matter of record, but few farmers have considered these findings germane to pasture management.

Often the phosphate load is so great it actually ties up other elements. Piling on even more phosphate and potassium can result in decreasing yield, not increasing it. It is a correct balance that accounts for highest sugar, highest yield, highest tonnage.

Suffice it to say an excess of phosphate can inhibit the zinc and copper uptake, with resultant debilitation to grazing animals. As André Voisin clearly illustrates in *Soil, Grass and Cancer*, zinc is necessary for nutrient absorption, meaning a benefit from rain and irrigation. This imbalance devastates grass quality and nutrition and yield. It was Voisin who discovered the Law of the Maximum. His book *Fertilization Application* makes the point that the Law of the Maximum is as valid as von Liebig's Law of the Minimum. The tag line of the syllogism is at once apparent. It says that too much compost or manure can increase the phosphate level to a point where they restrict the copper and zinc uptake.

Potassium was mentioned above. It always ties up boron and at a high level it can block out manganese. (For complete details on tie-ups and reverse effects of major nutrients, see *Eco-Farm: An Acres U.S.A. Primer.*)

Excessive nitrogen always blocks out copper, and at very high levels it also causes zinc deficiency. Excessive calcium blocks out everything.

Excesses prompt departure from the plane of chemistry and invite a roll-call of micronutrients.

High calcium may not be a problem in terms of cation exchange capacity, but the overload can play havoc with the livestock in the soil, namely, microorganisms. When the subterranean workforce is compromised, it cannot perform its anointed function, which is to take inorganic elements and make them organic for grass uptake. Trace nutrients, once complexed, will not assert themselves unless the calcium is returned to Albrecht's equilibrium.

Here a dilemma presents itself. It may be more expensive to use sulfur to cancel calcium than to find new sources for trace nutrients, which is precisely what old man Martin did by using ocean water with his manure fermentation described in Chapter 4.

Since it takes double the calcium excess to sulfur it out, the cost equation invites a look at many of the prospects eco-agriculture has available when an excess of calcium is compiled.

Not many grass farmers ever test. But they often go whole hog on using rock phosphate. Rock phosphate raises the calcium level as fast as it raises phosphate. That is why pastures seldom call for either lime or rock phosphate, but rather for manure judiciously applied and compost or compost tea.

The sequence becomes a mantra. The higher the calcium goes, the higher the pH goes, the harder it becomes to get on enough micronutrients needed for the appropriate levels.

Soil chemistry leaves a lot unsaid about the physical structure of the soil. For every point that calcium goes above 68 percent, it actually makes the effects of 1 percent magnesium on the physical structure of the pasture soil. There are pastures that would rip out in chunks as large as a broken paving stone. The word of choice here is imbalance, and the assembly of data asks us to recapitulate the cation balances, 65 to 70 percent calcium, up to 15 percent magnesium, a modest 5 percent potassium, and a modest sodium requirement. There is a pecking order in pasture nutrients. The strongest push the weaker ones out of the way.

Why the range of 60 to 70 percent of base saturation for calcium? It depends on whether the pasture is clay or sand. Magnesium needs to be 10 percent for high exchange soils, 20 percent on sandy soil. Sand needs more magnesium to hold it together. In other words, the pure mineral needs to be reduced in sand, enlarged in clay. Calcium and magnesium combined should never occupy more than 80 percent of the sites in terms of cation exchange capacity. Dead reckoning adjustments are never accomplished with a magic wand. For every 1 percent calcium is reduced, with nitrogen, magnesium increases 1 percent. Nitrogen will not control magnesium.

The soil readout will always be mysterious to those who do not pay out the price for the effort it takes to understand it. And yet the scheme is quite simple, although not as simple as watching grass grow under the assumption that if it greens up it will do the job.

Conventional wisdom associates the right potassium level (5 percent) with stock strength, winter hardiness, increased moisture uptake. An old Arkansas saying has it, "Potassium is the poor man's irrigation."

Grass never needs more than 5 percent potassium, whereas cotton and woody plants can stand an escalated level of 7 to 7.5 percent.

The desired range for sodium is 0.5 to 5 percent. Grass will eat salt if it is there or is provided — within the above limit. The salt lick is a poor substitute for sodium in its position as a cation. Many grazers are discovering ocean solids, or at least diluted ocean water, as a beneficial treatment, both to supply salt and trace minerals, sometimes as many as 90 or 92.

The use of Chilean nitrite for nitrogen and sodium is legendary. At sodium levels below 0.5 percent, barley won't grow. Sugar beets flounder. The cabbage family won't do well. And pasture grass merely abides the shortage as it does with most nutrients, always penalizing the cows.

Nature has her own checkmates. When calcium gets below 60 percent, sodium stops moving out of the soil. Above 60 percent and buffered with rainfall, sodium leaves.

The total exchange capacity can be said to represent all the positively charged nutrients, calcium, magnesium, potassium, sodium, iron, manganese, copper and zinc. The last several named are the "minute amounts" the pasture requires. In short, the Albrecht model considers the whole package.

Only about 28 percent of all farms take soil tests, probably because at least 70 percent of all farmers do not trust them. For those who deal with pasture, the lack of trust is even higher.

Many farmers let the signs and symptoms available in the pasture tell them what to do. This is quite possible if weeds and forbs are understood, if the living complex in the soil is treated kindly, if the grazing animals thrive. In the main, however, it is the type of sample, the language of the laboratory and the quality of the recommendations that preside. The Albrecht model depends on consistency of laboratory procedures and the computations that make figures flow into fertility amendments.

Availability of nutrients also figures. Lime for the pasture tends to break down in three-year increments, one-third each year. Rainfall or lack thereof speeds up or slows down effective plant use.

Exchangeable hydrogen is quite dependent on pH. Above pH 7, hydrogen will read zero. It takes a pH under 7 to permit exchangeable hydrogen in the soil system. The term here is *acidity*. At 10 to 15 percent exchangeable hydrogen, everything is in place to form the natural organic acids that will etch out phosphorus and potassium from insoluble materials.

Albrecht use to say "insoluble but available." In fact, one of his last papers defined the availability as compared to the full inventory in the soil, most of the nutrients being unavailable for plant use, "loaded up," "tied up." It takes hydrogen ions to form the mild organic acids to unlock, to give microorganisms a fighting chance at making organic nutrients out of inorganic stuff.

Salt concentrations are not often a pasture problem, "not always" being a hedged statement because all generalizations are false, perhaps even this one. In any case chlorides are easily leached out of the soil.

Boron is a different matter. Boron is always expressed in terms of pounds. For every pound of boron applied to a pasture, the only expected effect is 0.1 ppm, this under conditions of 30 inches of rainfall per annum. Every soil needs 0.8 ppm boron. Grass can do without, but not without consequences for the grazing animal. The ideal boron level for grass is 1.5 ppm. Boron is a disease fighter. It joins copper in fighting bone problems, fungal diseases and a miscellany defined as *Metabolic Aspects of Health*, a title still available from the Price-Pottenger Foundation. Without discussing pastures, that book takes it from soil to animal to human being, answering every question asked. It correlates perfectly with the Albrecht model and unlocks secrets of nutrition understood only by nature's animals for eons of time.

Boron has a just reputation for canceling out fire blight in Russian olive trees. Of more interest is its role in nitrogen conversion in the soil. Anyone who grows grass for seed discovers the boron mandate soon enough.

Iron should be 200 ppm. Usually it is only 2 ppm. If the grass is still green, then roots are still moving down or pulling iron out of water. Iron is almost always available in the subsoil, always answering the call of a low pH.

Manganese should be 40 ppm. Even with lots of money, it takes two years to bring manganese to an appropriate level.

Copper at 0.7 ppm needs to be a minimum of 2.

Zinc, often 0.3, needs to be 6 ppm. When grass still needs water after a heavy rain, it needs zinc. Zinc is necessary for water absorption. An application of, say, 36 percent zinc sulfate will cancel out wet immediately. At pH 7 that 30 pounds of zinc sulfate would get the above-described farmer back to the right zinc level in two years. At pH 8, probably two or three applications would be necessary, these over as many years to bring the zinc level back into equilibrium.

Without the major parts of the Albrecht model in place, it might take years to achieve suitable availability for the major micronutrients.

It should be clear by now that each pasture has to be addressed according to the soil audit readout, each level dictating a new answer. As the above is considered, there arrives the ultimate questions — when, what, how? Ammonium sulfate works well with the microorganisms in the soil. A dry season application, possibly 200 to 250 pounds per acre will get the grass through the winter if only conventional nitrogen is available. A spring application may be indicated, slow release is the key even when a dry year follows.

A strictly organic pasture needs its compost or compost tea, even protein meal in late winter. A microbial stimulant will also help the system generate nitrogen. Twice a year is an oft-recommended procedure.

Copper has become a debate topic because of its identification with the Mad Cow syndrome, its lack of uptake being the issue, this according to Mark Purdey of England as backed by Cambridge University. Its uptake, not absence, is seemingly the issue when Phosmet is used to battle grubs. Even when available in pasture grass, uptake can be a problem under conditions of magnesium excess described by Voisin in *Soil, Grass and Cancer.* Copper, like zinc, is a disease fighter — "take all" disease is a copper deficiency. Eliminate the deficiency and you eliminate the disease. Appropriate copper levels in the grass mean an absence of crop diseases and a likely elimination of animal vet problems. Copper is necessary for protein conversion, strong arteries and veins, prevention of aneurisms. Turkey manure almost always suggests a copper payload because of copper supplementation to prevent aneurisms.

Do micronutrients take care of themselves when cations are in balance and major nutrient requirements adequate? Generally, yes! — if they exist in the soil. But as George H. Earp-Thomas proved early in the last century, cobalt is a non-resident of almost all U.S. soils. In most cases the micronutrients are either absent or complexed — absent being the most likely.

The pasture both demands and delivers. That is why foliars are of maximum importance in terms of micronutrients. Sprayed on the grass, they are used as needed, the balance being remaindered to the soil in the area of the roots for future uptake. In that manner, foliars in fact help feed the microbes in the soil solution around the root hairs.

Just the same, foliars are best used as an interim remedy, a crutch pending getting things right. Grass promptly says, "thank you," but it still asks for the soil's blessing.

Manganese is the second key to plant strength. It is the key to seed set. It should be kept to 40 ppm as a minimum. A 250-pound fix of manganese sulfate will do the job at once for pasture grasses, always to become that pH of 8. That iron which was mentioned earlier now needs attention even in the absence of yellow grass. Any time the iron is higher than the manganese in the soil, the manganese ionizes the iron and it is no longer available for the plant.

Ferrous sulfate, 400 pounds of 21 percent, is indicated because it will not kill plants, ferric sulfate will. If ferrous sulfate is black, it is worthless. If the product is white or blue-green in color, it works. If the material is rust colored, it does not have enough water molecules in the chain to work, and it will not build iron in the soil. The white and blue-green is appropriate even under organic rules.

Ferrous sulfate drives calcium out of the soil, as Rudolf Ozolins suggested in an early issue of *Acres U.S.A.* Western Kansas used it effectively to deal with their high-calcareous soils, usually at a ton per acre. There is now available an iron sludge, an organic product treated with microorganisms that is somewhat affordable.

As these lines are set down, there is a concerted effort going forward to buy up all natural deposits in North America, this according to Neal Kinsey. "When that happens, the price for micronutrients will double, even more," Kinsey told this writer. College research is bound to follow, after which everyone will be told about micronutrients, essentially ratifying what *Acres U.S.A.* has been saying all along.

Without waiting for such a turn of events, a note must be added on boron. To get boron above 0.8, 10 pounds of 11 percent borax will raise the level 0.77 to 0.87 on a 25 exchange capacity soil in one year. Lower exchanges won't comply with this norm.

The Albrecht model makes assumptions that seem to have taken their place as "settled" science. The old pie chart that divides soil systems three ways, physical, chemical and biological, is confirmed, furbished, and refurbished. Chemical analyses do not tower above biological, and certainly tilth confirms the efficiency of the Albrecht treatment. Crops confirm the model, and grass capable of producing grass-fed beef positively ratify the concept that grass, indeed, is "the forgiveness of nature, her constant benediction."

Tests on coastal Bermuda grass stack up like cordwood. Usually the exchange capacities of the sandy soils involved are low, thus remedies would have to be worked incrementally. Now the Albrecht percentages become malleable, pushed this way and that, to accommodate the small maneuvering space the cation exchange capacity affords. Otherwise the rules remain the same. With a low holding capacity, it no longer remains possible to feed the soil in order to feed the plant. The key becomes incremental treatment.

6

Holistic Resource Management

In Holistic Resource Management (HRM) ranching, carbon dioxide does not escape into the air. It quickly gets reabsorbed.

Texas A&M has added to the knowledge of this subject. On perennial grasses, some of the roots die off every year. It was a revelation that had not occurred to the investigators even though the source of carbon dioxide remained unanswered.

Rhodesian biologist Allan Savory left the land of his birth and settled in Albuquerque, where he immersed himself in the business of counseling ranchers how to manage cattle on fragile soils. The insight he brought was African, developed by the animals, much as the bison once developed a system for maintenance of America's vast grasslands.

Three things were essential to the holistic model. This meant that every former government operating resource, whatever, needed to apply three benchmark considerations and more.

First, they had to determine the whole of what was being managed. This cancelled out the idea that one had simply to manage a ranch or a cow pasture or a forest. The whole had to involve the human element. For the cattle operation it meant involvement of the entire family, even an extended family.

The land units require their own analysis. The finances have to be involved in the consideration. Failure to consider the whole always results in a breakdown. The whole that parades through the mind is always a part of a greater whole, this on up to the universe itself.

The second essential requirement is the human element associated with a goal. It means a signal failure when the goal is incomplete. American agriculture seems to have a common goal — abundant production. Allan Savory would call such a goal a non-goal. It, he says, leads to trouble. The American record seems to bear him out. Production has been prodigious, but the damage to the environment has been brutal to families, to land, to rural life, to the total environment.

The goal should be quality of life, production that will sustain that quality of life gained from beef, timber, grass, and the conversion of solid revenue into opaque production. The educational aspects of that life are a requirement.

The last element of the HRM goal sequence is a landscape description. This description pertains to 1,000 years from the present based on production and quality of life sustained from that land.

Thus a philosophy presents itself as a keystone element in the HRM model. I asked Savory what he meant. "When the landscape description is devised, we do it in terms of a general description of what the land will look like specifically in terms of four functions: How nutrients cycle, how water cycles, how the successional process functions, and what it would look like idealistically on that land and how energy flow occurs. What sort of energy flow do we need on that land? When we have this comprehensive goal — and we have to have one for every farm — we move into the third potential model, namely how to achieve the objective."

Now the thought model becomes engaged. Savory stresses that this cannot be done on computers. Holism makes the point that many things are not quantifiable. Human values cannot be reduced to an electronic blip on a screen. As a consequence, having a whole and a goal selected cannot be governed with the interdisciplinary approach.

One goal is to stop desertification in its tracks.

HRM people make the case that fully half of the farm exodus could have been stopped over the past several decades simply by invoking the HRM model. The mechanism is grass, livestock, short course instruction, in short, the model.

André Voisin, the author of *Soil, Grass & Cancer*, has been credited for at least one leg of the HRM model. When Allan Savory came to the United States, he brought along other elements of the HRM thinking, much of which had still to be fleshed out. Savory was born in Rhodesia, now Zimbabwe. He has described himself as a fanatical conservationist or environmentalist. Like Jeremy Rifkin, he believed all cattle should be removed from the land. He would have nothing to do with them because obviously they were raping the land. He tried to solve the problem from a wildlife perspective. He found that there were many situations which could not be solved with game alone. In the course of working with that problem he nailed down no less than three discoveries.

1. There were two types of environment with totally different decay processes. For any environment to flourish you have to have birth, death and consistent turnover. With any population the decay process is as important as the life process.

"We had assumed that the decay process was essentially the same on all ranges and watersheds." Savory finally admitted "they were not."

That is why he came up with the terms *brittle* and *non-brittle*. These are not to be confused with *fragile*. In fact, it has nothing to do with fragility at all.

Savory calls non-brittle areas of the world such as the eastern part of the United States, Europe, New Zealand, and tropical Africa, due extension of this concept to all appropriate parts of planet earth. The decay process in these areas is largely biological. It is rapid. It tends to take place from the base, from near the soil on, say, a dead tree or grass plant.

Most of planet Earth qualifies as brittle environment. Here the opposite is true. The decay process is slow in the extreme. It may take several years and tends to be from the top of the plants down, via physical wearing of chemical oxidizing materials. The significance of this, notes Savory, becomes great when the decay process is slow — from the top down. Major watersheds rely on grass for stability.

2. Grass grows from basal growth points close to the soil surface. Grass requires light. If light is blocked by a slow decay process, the plants tend to weaken, kill themselves, grow further apart! The discovery was simple once it had been made.

3. Another discovery related to the environment was equally apparent. If you rested the land, a successional process would take place. In the non-brittle environment of the world, this was true, Savory observed. Non-brittle acres that are rested invite successional processes that move even in the most extraordinary situations of deep slopes. The brittle environments of the planet serve up a different picture. On smooth surfaces the decay process is initiated with extreme difficulty. Very often there is an algae or lichen phase level that stays on for years, even centuries. Very steep slopes cancel out even that level. The great canyons of the west provide a good example. Some of these areas have been rested for thousands of years, yet they still actively erode. The slope is simply too steep for such a brittle environment.

Rainfall is not the sole answer. The total miscellany of that climate comes closer to defining brittle and non-brittle. "We used to say that the arid and semi-arid areas were the areas that were deteriorating more, but this was making no sense to me in Africa because some of the areas I was working in had 50 to 80 inches of rainfall and they had slow chemical decay processes and very bad deterioration taking place in conventional agricultural practices," Savory explains.

Certain conclusions flow naturally from these observations. Low rainfall on non-brittle end of the scale and also in the brittle, but what really governs is the distribution of atmospheric moisture throughout the year. Otherwise "we find that in a non-brittle environment, if the rainfall is low, the distribution of atmospheric moisture is favorable for the buildup and sustaining of microorganism populations. On the brittle end of the scale, the distribution of atmospheric moisture is such that though there may be high amounts of moisture at times, there are brittle weeks in parts of the year and atmospheric moisture is low." As a consequence, most biological populations are drastically reduced. At the end of such a year the decay process is slow.

These grasslands often endured drought, and yet they achieved climax crops in the wake of herd traffic, fertilization and tillage.

The bison swept over those grasslands as tight herds, grazing and dropping urine and manure. They were on the move while they obeyed their instincts to increase and multiply. Always, they moved on before all the forage was consumed. Usually they would eat one-third or slightly more of the available grass. A lot of green leaf surface remained as the herd moved on. The manure gifted to the soil was broken down as a consequence of traffic and compact grazing. Stomped into the soil and broken down and soaked in, these materials released carbon dioxide.

Carbon dioxide is heavier than air. It stays in the canopy of the plant to a remarkable degree. The stomata on a plant leaf quickly drink in the carbon dioxide as nourishment. Once satisfied, these pores close. When they open they do not need to stay open very long in the presence of a carbon dioxide flush.

Plants transpire into the air at least 99 percent of the water taken from the soil, the agency of that transpiration being the open stomata. When the presence of an ample supply of carbon dioxide gives the plant its fill, the stomata closes and loss of water from the soil is rationed. That is why one farmer will experience drought while at the same time a neighbor has no drought. With a decay system operative, the steady release of carbon dioxide closes down excess transpiration, hence a soil that endures a shortfall of rain and still produces a crop, retains its moisture and contributes little when dust clouds form.

Greenhouse growers discovered this connection decades ago. They noted that photosynthesis would shut down by 10:30 a.m., when carbon dioxide ran out. Clandestine marijuana growers who often maintain production facilities in basements, attics and warehouses know they have to fertilize the air.

The problem with piping in carbon dioxide is often a disturbed balance with other nutrients. The plant cannot utilize excess carbon dioxide if the mineral support is not available. When the buffalo dropped all that manure and stomped it into the ground later, soil microbes and roots — even dung beetles — were being fed in abundance.

The lesson became obvious to those who observed and connected with what Allan Savory was saying. Cows grazing intensively in a paddock deposit a lot of manure. A lot of carbon dioxide is given off while the manure is still on the ground.

Most readers know that farmers export two bushels of soil in erosion for every bushel of corn produced. To increase production on American rangelands, we have poisoned, chained, plowed and reseeded them, and yet over 223 million acres have experienced severe desertification and approximately twice as many acres are threatened. To answer these problems, we have engaged ourselves in almost two decades of intensive reporting, snagging lessons from any available font of knowledge, yesterday's and today's, always probing for insight and results. Not until we ran into Allan Savory did we find anything entirely different. Savory calls his school of thinking Holistic Resource Management. He discusses his thought model in this interview.

Who is Allan Savory? He is an ex-Rhodesian wildlife biologist and racial egalitarian. Two of his basic contentions are that overgrazing is not a function of stock numbers, and he holds that cattle, sheep, goats and horses may be the cheapest, most natural and only realistic tool for restoring lands devastated by cattle, sheep, goats and horses. He calls himself a self-directed learner. With a degree in zoology, he "was too poor and Africa was too far from America" to do other than train himself. He had hoped to get an American Ph.D.

While working as a wildlife biologist in a remote area of Zambia in the early 1950s, Savory was struck by the contrast between the lush and productive areas that teemed with herds of game and areas nearby which were deteriorating rapidly with little or no game. To understand why, he closely observed the effects the animals had on the land. As they passed over an area, their hoofs broke up and loosened crusted soils and trampled down old plant parts thereby creating mulch and an ideal seedbed. At times the area appeared devastated but would recover dramatically in the next growing season. The extent to which the animals disturbed the land was directly related to the presence of predators which kept the herds bunched and excited.

Before man interfered in areas like these by removing predators or disturbing natural movement patterns, similar areas throughout the world had supported large herding wildlife populations for millions of years. The plants, soils, herding prey and predators had in fact evolved together and needed each other for their own health. This was the first missing key.

In those areas that had not evolved with herding animals, such as certain grasslands and jungles, the rainfall tended to be reliable and the breakdown of old plant material very rapid. Because of this, these areas were highly productive and contained a great many plant and animal species, even when not disturbed.

In the areas where herding animals had historically been present, rainfall tended to be erratic and the breakdown of old plant material very slow. Because of this, these areas were less productive and had fewer species where undisturbed. However, as the herding animals trampled the old material into the soil, they sped the breakdown process and thereby increased productivity. The degree of productivity and diversity was dependent on the presence of the herding animals and their predators and the consequent periodic disturbance.

From these observations, Savory realized that the way an environment reacted to herding animals was dependent on the climate and decay process present. What he termed "brittle environments" had erratic rainfall and a slow decay process. To be productive and stable these environments needed the herding animals and their predators. What he termed "non brittle" environments had reliable rainfall and a rapid decay process. Productivity and stability were not dependent on herding animals and their predators. This was the second missing key.

Meanwhile, a French researcher, André Voisin, discovered that overgrazing did not reflect animal numbers, as commonly believed, but the time individual animals spent in one area. A single cow would overgraze some plants if allowed to bite them repeatedly as they grew. Huge herds would not overgraze if they moved after the first bite and didn't return until plants recovered. Thus *timing*, not numbers, governed overgrazing. This was the third missing key.

In 1964 he left government service to operate a sugar farm. In the fullness of time he operated his own farm, two ranches and a private game reserve. He served as a consultant in over five countries while he was developing the insight that would one day become holistic resource management. He also spent eight years in Rhodesia's Parliament, and then he came to America as a political refugee, having been forced into exile by the Smith regime. "I was leading the political opposition of the moderate whites in the Smith government," Savory told me. Although he did not serve under Mugabe in what is now Zimbabwe, he has a good relationship and recently had five workers over for an intensive training program.

Savory's discourses are so chock-full of insight, of holisms, if you will, it is not possible to handle everything he says in a short report.

When I had completed taping, Savory continued: "People in West Africa are working on alley cropping. This is simply rows of leguminous trees, with the crops interspaced between them. It is very sound, very promising — a wonderful work, and we are encouraging that a lot." Is this holistic or common sense? Savory: "It isn't holistic because as good as it is, it does not take into account the market, the family, the matter of chemical dependence, whether there's just been a divorce — and politics and economics." Or as *Acres U.S.A.* sometimes puts it, "to be economical, agriculture has to be ecological" — and now we might add, "holistic."

Q. The purpose of this interview is to field some of your comments and ideas and to get an explanation of what you are doing in eco-agriculture. As we understand it, you are conducting instruction courses outside of academia and taking a lead in ecologically sound land management?

A. Yes. But we have a lot of people from universities in our courses. So we tend to instruct academia outside of academia.

Q. What is the basis for your courses of instruction?

A. You are aware that we have enormous problems facing us world-wide in many fields, and our economy is going haywire. Agriculture is obviously unsustainable. Most informed people are increasingly aware of a lot of the damage being done to the environment by conventional agriculture.

Q. We seem to have come full circle from the old mercantile days, when the king could send out his corporations to plunder the earth. Now we are plundering the earth in a lot of respects, including wasting away the natural resources?

A. That is correct. If you look at these many, many problems, deforestation, desertification, the deterioration of our watersheds and in our arid and semi-arid areas — if you look at all of these problems, you have to note that they are getting worse. They are not getting better. There is no country in the world that is on top of these ecological problems.

Q. Least of all the United States?

A. Yes. Least of all the United States. The rate of desertification in New Mexico, Arizona, west Texas, is quite frankly faster than anything I have ever experienced, and I come from Africa. So you know the rate of these things happening is pretty great here. These are the sort of things that are happening fast here. Up to

400,000 farming families a year are leaving the land. This is blamed on over-production and low price. But, can you name any country in the world where the farmers have enjoyed consistently higher prices than in America and in particular where the farmers in the world have enjoyed such low price input item, tractors, fertilizers, chemicals, etc.? Price is not the problem.

A Model, Not a System

Wes Jackson of the Land Institute, in Salina, Kansas, once said that if we had the working manual of a corn plant, it would fill the shelves of a sizeable library. Allan Savory would agree, and this is the reason for being of Holistic Resource Management. Savory rejects the idea of systems now that experience and observation have revealed a hierarchy of relationships that go back to the total earth itself.

There is no country on planet earth that is on top of its ecological problems, says Savory, and this permits us to take a quite different look at grass, problems of the pasture and of farms, ranches, forests, the total landscape and the objectives of the people responsible for reacquiring the values.

What happens to grass happens to the land, and the mindset that assigns fault to prices, production and the people in Washington fails the test of reasonability — when a holistic view is considered.

Holistic Resource Management is a model, not a system. This model has assembled no less than three concepts, one of which was mentioned above and is reiterated among the three.

André Voisin discovered that the time animals spend at grass was more important than other factors. Savory discovered the other two factors. He discovered land degradation could be stopped at very low cost. And he found that the interdisciplinary approach failed to answer the vast complexity of causes that conventional wisdom tried to understand. In other words he sought to merge agronomy, wildlife and an inventory of considerations that brought family aspirations into the box and projected that future not for a year but for 1,000 years.

No two farms are alike. Even the individual attitudes of proprietors shade the decisions and the results.

Therefore the model, not the system, decreed answers that are to be as malleable as hammered gold.

"We found we could diagnose problems like a grasshopper outbreak and determine the best way to overcome it," Allan Savory said. A major key was seated in grass management. Instead of squandering millions of dollars in quick fixes, the model created benefits to farmers, foresters, waterway operators, fisheries, etc. Economic modeling covered spinoffs.

Controlled Grazing

Now the issue of controlled grazing asks for consideration. This is safe in the east, in New Zealand, in non-brittle areas. But that same thinking inserted into brittle areas — meaning most of the Western states — "It can be extremely dangerous to the rancher," Savory opines.

No single factor seems to be the answer. Remove the animals, don't remove the animals, do this, do that, none or all demand scrutiny in terms of the holistic approach.

HRM holds that the arguments between environmentalists and ranchers exhibit a good example of options on both sides that are 100 percent right and 100 percent wrong, depending on the land unit involved.

In unraveling the scope of this HRM insight, I returned in my mind's eye to western Kansas. The rainfall was 10 to 12 inches per annum. By the time you get to Topeka, rainfall could be 27 inches in a good year. I remember old buffalo wallows left over from frontier days. The prairie grass was almost flawless in undisturbed steppes of Volga-German farmers. It wasn't a monoculture. That short grass was almost pristine.

In terms of HRM, this was a brittle environment. It was not as brittle as New Mexico.

The tools indicated no doubt called for what HRM calls for a "disturbance." These acres of grass were maintained by herds of bison that disturbed the land.

Buffalo, elk, deer, antelope, all must have figured. Periodic fires also helped. The legacy of that formative period permitted farmers to create a great breadbasket at the expense of grass-building. Such an area can be rested, resulting in an apparent rebirth of production. Unfortunately the decline comes on, slow and gradual. In the case of Kansas, the golden days were terminated by dust storms. Weather cycles and irrigation from the Ogallala Aquifer now permit continued rape of the soil with the help of salt fertilizers and/or rescue chemistry.

When the decay process slows nature's plan, the resultant interference with the water cycle has an impact well into the Gulf of Mexico. The nutrient cycles are also affected dramatically. Such an area suggests a thought model based on the whole. The resources call for management of the entity.

The ramifications suggested by the HRM model can and have filled books. Here I would like to consider the nemesis of the 1930s, grasshoppers. Here is the suggestion Savory made, using the HRM model in a diagnostic mode. His references were to outbreaks in Wyoming, Idaho, and Montana. The partly brittle environment is often subjected to partial rest. This means cattle on the land are scattered widely to protect the soil from trampling. Grazing cattle in this manner tends to open up the space between plants. Also, if plants are overgrazed, there is a tendency to open up the space between the plants. When the space is opened, more overgrazing results. If the decay process is slow, as it is in a brittle environment, then there is an accumulation of oxidizing material. Often the Flint Hills or western Kansas or Oklahoma or Texas rancher resorts to fire. This introduces a last factor in opening up the space between plants.

Open space means added breeding ground for hoppers and survival of eggs and insects in the nymph stage. Those organisms can be expected to increase dramatically.

These are precisely the conditions that accommodate egg survival. An outbreak of grasshoppers often result. Running cattle in a way that will close the space between plants is a tool. Sprays are excellent in fiction, mediocre in fact. Resistance will build up and a super-insect will result. Egg-laying rate will not be diminished. Moreover, there is a fundamental problem with chemicals of organic synthesis. When you kill the predator you also kill prey. The prey always recovers first. If the breeding sites are not removed, the next outbreak will tower over the earlier one.

Application of the HRM model to each land unit never relies on a single-factor analysis. Always an issue is the matter of goals — farm management. All figure. If single-factor analysis is allowed for a moment, then the spacing between plants calls up the available tools, some mechanical, but mostly high animal impact. Many hoofs would have to rough up these surfaces. Savory uses the term *tool* to mean all tools available to mankind. It is the concept of animal impact that invites the most attention. Grazing but not overgrazing ranks high as a possibility. Fire is not an option. Rest for the pasture would exacerbate, not solve the problem.

Grazing and animal impact is now an expanding trend on brittle acres, always answering a set of guidelines.

There was a time when predators kept the wildlife herd in a tight bunch. The animals grazed and moved on, never organizing, always leaving fertility behind. Fences now serve as predators in holding together the herd that has been assigned the business of churning brittle ground, leaving enough leaf area for continued photosynthesis and recovery.

Malcolm Beck

Malcolm Beck, the San Antonio composter, has delved into the texts of an earlier era. "They knew a lot of this stuff," he said, "but they never tied it all together." Seekers who have morphed their philosophical insight into a working reality, especially west of the Isohyet, call themselves Holistic Resource Management. The results they have conjured up out of fragile soils using lessons from exotic places like the Serengeti now tantalize conservationists who are tormented by the march of desertification.

Impact has many dimensions. The one easily overlooked is the new root production that attends fertilization, renewal after impact, carbohydrate production and renewed photosynthesis. Old roots decay trapping carbon dioxide. Decay nurtures earth worms, microorganisms and a raft of subterranean life.

We thus return to the proposition that attracts so much attention these days, the propensity of grass to sequester carbon. Trees are often cited as a sure trap for carbon dioxide, but trees do not require carbon dioxide to hover near earth's surface, a position required if carbon is to survive grass and crops.

Simply stated, tree roots die only when the tree becomes food for termites. By the time carbon dioxide arrives under tree leaves, it has dissipated. Pores on tree leaves cannot capture carbon dioxide on par with grass.

Malcolm Beck relates a case report of a farmer named Bobby Sparks. In performing research work for Joe Bradford of Kika de la Garza Agricultural Research Center, Weslaco, Texas, Bobby measured his soil, .8 percent when he started. In eight years he increased his organic matter to 1.6 percent. The increase was 0.1 percent each year. This is the amount each acre needs to increase its organic matter if global warming is to be cooled.

If the working model of a corn plant would fill a library, then nature's variables would likely fill all the libraries of the world, leaving no room for history or literature. To strive for a holistic view is both ambitious and humbling.

Beck recently told me a story that suggests grass may be more of a miracle than we think. He told of a dry lake that drained California acres being farmed mechanically. Children hunting with .22 rifles found deformed animals in the lake. This prompted scientists to test the sediment. It was found to be overloaded with 700 ppm selenium. Selenium is a nutrient up to 4 ppm. Then it becomes toxic. At 700 ppm the substance is highly toxic.

The Bureau of Land Management wanted to cover the lake bed with two feet of soil. This, of course, would have done no good whatsoever. Then the suggestion was made that compost might help solve the problem. No one knew how.

The only material available was pulp and peels from an orange juice operation. It was repeatedly sheet composted, one 18-wheeler after another delivering the materials.

Tests revealed the selenium quickly going down. No one knew where it went. So they pop up a dome of sorts, captured air and measured it. They found that the microbes were combining selenium with other elements that constituted an inert gas that dissipated over planet earth.

We understand very little about nutrients distributed by air, and we know even less about still unnamed life forms. Beck put it this way while we discussed holistic resource management, insects, carbon and grass.

"As long as the selenium concentration was very toxic, the microbes seemed to fall to the task of working it. As the level got to four ppm, they went off duty." Nature put those microbes to work saying, *This stuff is bad, you all work on it in a hurry.* On the chart, Beck said, continual composting for 10 years would reduce the toxic overload to a nutrient level of four ppm.

"When you go to studying nature, you find that there is a master design. It is beautiful. It is perfect. In my insect program, I show that every insect on earth is designed to do something for nature. Every living thing on earth is programmed to do what it does and be what it is, except human beings. Human beings have a free will. We can be inert, we can be stupid, we can kill each other, we can live in harmony. Every creature was designed for a purpose, to show us something, give us something, tell us something. Study, and all of a sudden nature opens her books to you."

HRM has opened the book on grass management. They asked Thomas A. Edison what he considered to be the world's greatest invention. The inventor paused a moment and answered, "A blade of grass!"

As far as the plant is concerned, CO_2 could go up to 2,700 ppm. But the plant can't prosper unless the roots are balanced with the proper nutrients. These findings were made before the advent of chemical fertilizers. In fact the researchers were using organic fertilizers, and consequently the roots were balanced out. The old norms juxtaposed to chemical fertilizers cannot work. The role of the stomata didn't make it into the literature in any big way until NASA got into the act. That agency published its findings in *Discover* magazine after space exploration got underway. Farm magazines have still to pick up on the stomata connection, *Acres U.S.A.* excepted.

NASA was studying global warming. This work brought into focus the cooling effect of green plants, grass included.

Dan Carlson's work with sound and nutrients — both dependent on open stomata — is almost a lone exception to the veil of oversight.

The connection now becomes clear. NASA found that the more concentrated the carbon dioxide in the air, the longer the stomata stays shut, and the less transpiration. Transpiration has a cooling effect. Without transpiration, the leaf heats up — as much as two degrees beyond the ambient temperature. NASA concluded that this too contributes to global warming. Absent in the NASA report was the quite obvious fact that a closed stomata means less water taken out of the soil. The last line of this syllogism is simply that transpiration contributes to global cooling.

Unfortunately the syllogism does not go far enough. If green grass and its companion plants covered the earth, the sun's energy would find absorption for carbohydrate production. Bare soil — parking lots, highways, cities and plowed ground — absorb solar energy and heat the soil.

There are approximately 455 million acres of farm land. There are some 520 million acres of pasture. If the farm land or crop acres could increase their organic matter content a tenth of a percent a year this would affect the carbon dioxide put into the air by burning fossil fuels. The seekers who proffered this information did not know how to do this.

The holistic resource management people do!

Dick Richardson, a University of Texas professor, has made the statement that grass sequesters more carbon than any other plant. I encountered this information before, while writing *The Carbon Connection* and *The Carbon Cycle*. This carbon sequestering propensity is precisely what brings Holistic Resource Management into focus.

7

Grass, the Forgiveness of Nature

If seekers would like to turn agriculture inside out and have done with the synthetics of N, P and K and toxic residue chemistry, they have an ally in Wes Jackson of the Land Institute in Salina, Kansas. A product of the Kansas University system, Wes moved on to California, and after a while he abandoned a tenured university position to challenge the principles most of modern agriculture lives by. His focal point was grass, meaning the grand diversity of plants that make up the natural cover of the Great Plains. His farm near the Smokey Hills River is privately supported, and his aim is to put grass in tune with a sustainable nature. "His mind bobs along like a prairie jack rabbit," a *Wall Street Journal* writer noted, "from soil erosion to the philosophy of Alfred North Whitehead."

The quoted parts of this chapter have been transcribed from my own Socratic talks with Jackson as well as abstracted from his testament, *No Man Apart.*

"Once we started farming and expanding our scale to the field level, we became a species out of context," Jackson contends. "We no longer gathered and hunted. We changed the face of the Earth." As a consequence, there has been a loss of ecological capital "and the expectation is that it is as sustainable as the nature we replace."

The prairie is climax territory. Jackson reminds all those who will listen that there are two things about the native grasslands of the prairie. First, the human being spent a long time being shaped by the African grasslands. The counterpart is the prairie of North America. Jackson's analysis comes about as close to an absolute statement as a scientist ever gets. "Prairies that are live provide the standards against which future agriculturalists will judge their practices." Jackson points out the fact that the prairie pays its bills regularly. The nutrients are cycled, there is an accumulation of ecological capital. The prairie has a water management system. It has a soil texture that is accommodated to a sustainable growth. "Sustainable yields," Jackson adds to the summary above, "and those ecological principals found in that prairie are precisely the same ecological principals that need to be utilized as we bring about this paradigm shift from a destructive form of agriculture into a sustainable agriculture."

When Jackson's words sink in, they seem to flash ancient scenes across the theater of the mind. Those who are pushing 80 or beyond can recall the dust bowl days of the 1930s when nature served up a most severe lesson. It was a case of reward and punishment in one time frame, all in one geographical area.

The prairie as a plowed experience was more devastated than the prairie that was overgrazed.

The prairie features a diversity of perennials, whereas the adjacent wheat field features annuals in a monoculture.

This serves up a question that has not been answered, probably because no one of Jackson's stature has ever asked it. Can perennials and high yield go together? Can an agriculture be developed that features enough roots to hold the soil? Can it go beyond the benediction of grass and grazing to achieve a seed crop on par with the soil-destroying crops now state of the art?

To ask these many questions is to suggest there are answers. The harvest of forage is a given and quite sustainable when graziers invoke the lessons made a matter of record over the past half century.

Almost all the erosion in the United States is a consequence of the human penchant to harvest seeds. Jackson's new look at the prairie asks for an encyclopedic review of all the plants nature's evolution has accounted for. He calls that part of a very long-term study polyculture of perennials that will outyield a monoculture of annuals. Can an ecosystem be made to sponsor its own nitrogen fertility? Finally, can such a system structure become an enzyme system capable of countering insects, pathogens, weeds? Can nature's grasses be managed?

This meant a literature search, a literal count of perennials' yields, an assembly of available perennials. The original countdown identified some 300 heirlooms as opposed to woody perennials, all planted in 5-meter-long runs. Six species and six genera opened the door to this new look at nature. And then it occurred to Jackson's team that "we may be missing something."

An order went out to the Plant Introduction Center, Pullman, Washington, asking for the relatives of those six species. They sent back the world's collection, 4,300 total, and "we planted them all," Jackson said.

From those many plantings came the main candidates — giant wild rye, a plant so tough it grows on Siberian sand dunes — was one. The Mongols used it for food, and it is a close relative to a plant the Vikings brought to Vineland. They called this one lint grass. Jackson observed that the Viking version was inferior to the Nagel strain.

One species that commanded the attention of the Land Institute team was Eastern gamagrass, a relative of corn and a popular resident of grazing paddocks from Texas to Canada. Eastern gamagrass checked in at 27 percent protein, three times that of corn but still not in the ball park with Schnabel's wheat as a grass grown on special eastern soil. Eastern gama had twice the protein of the usual wheat crop. It tested 1.8 times higher in methamine than corn. It is an important amino acid.

Wes Jackson explained, "We have a mutation that forms the male part of the flower into a female. This has increased the number of seeds produced by an individual plant twentyfold, and the field by weight some five-fold."

The product of this breeding will start, flower and produce seeds for weeks and weeks. Results suggest an 800-pound-per-acre yield.

The next step was selection for the purpose of determinant flowering, the purpose being flowering and seed setting all at once. The objective being 2,000 to 3,000 pounds per acre. Corn often comes in at 5,000 to 6,000 pounds per acre.

There are other considerations. The gama cited above fixes some nitrogen in the rhizosphere of its roots.

As the Land Institute emerged from its inventory phase, other plants were considered, the Illinois bundleflower being one. This one delivered 34 percent protein from the start. Cracked like soybean, it proved 85 percent digestible. As a perennial it fixed nitrogen. This seems to be the reason for being of many grasses, fixing nitrogen. It can be computed that the business of supplying nitrogen cost at least twice as much fossil fuel energy as operating the tractor on a row crop plan.

The Jackson experiments, calculated to run at least 50 years, perhaps 100, is more than a side departure in this assembly of valid grazier information. It states with promising finality the proposition that only a return to soil-building biennial crops can recapture the value of the land, as grass constructs enduring roots, gives up a measure of that root structure for carbon dioxide release, and even holds promise for seed protein production, with a protein content superior to the soil wasting monocultures that have reduced organic matter to less than 1 percent for most of America's acres.

"One plant we have considered," said Jackson, "is the Maximilian, or sunflower. That plant is allelopathic, which is to say its roots exude a natural chemical that acts as an herbicide. During an interview several years ago, Jackson could point to a plot kept weed free based on the Albrecht dictum that plants in tune with exchangeable nutrients have the capacity for constructing their own hormone and enzyme receptors necessary for battling weed, fungal and bacterial crop destroyers, and weed proliferation.

Many graziers think of their grassland swards as a monoculture, a good stand of bluegrass, brome, or other genus so specific they have a common name, yet defy transfer from one latitude to the next, having evolved into harmony with the environment.

The species in the short grass prairie outdistance the imagination. One area near the University of Nebraska — one square mile — served up approximately 270 vascular plant species, according to Wes Jackson. "If you were to take the DNA for a square mile of prairie grasses, the working manual would fill more than all the libraries of the world." Hyperbole or not, such a statement makes a point, one Jerry Brunetti ratifies in his section of this book.

"A prairie is information rich," noted Wes Jackson. "A corn field is information poor."

These few remarks suggest that scientists are going into cloning and genetic engineering with insufficient knowledge.

The USDA tells us that there are 1,775 weeds in the United States. Some weeds are grasses, and possibly half the grasses and weeds are medicinal botanicals. The point can never be stated too much that the grazing animal knows how to read the benefits and deficits of forage, and has been in possession of this knowledge for centuries before reductionist science ever started scratching the surface.

For instance two unlikely kissing cousins in the plants listed by Jackson and associates are Johnsongrass and grain sorghum. Grain sorghum is a perennial, but it winter kills. Johnsongrass is a relative, generally considered a noxious weed. Land Institute has crossed sorghum with Johnsongrass, but of that marriage came a few individuals that proved winter hardy. F-1 times F-1 derived the F-2 generation.

It takes an Alaska-type winter to prove a point, and so the work goes on. Man proposes and nature disposes. Experimental works put into focus the awesome job nature has done through eons of time. Yields with winter hardiness built in suggests an attainable objective, on paper a hybrid derivative of Johnsongrass with a tetraploid (four sets of chromosomes) times the tetraploid sorghum. Most sorghum is diploid — two sets of chromosomes. Always, the objective is a perennial that will kick off a massive breeding program.

Range Concepts

Henry Turney, the rancher who brought pasture and range classes to the Texas University system, has left his course outline as a book entitled *Texas Range and Pastures.* He might as well have said U.S. range and pasture, as most of the grasses he cites and locates in their preferred Texas habitats do not confine themselves to the Lone Star State. Turney's chief grasses backbone any list, if indeed a list is of value. They are grasses, each claiming its own space and clime — Indiangrass, wintergrass — sometimes Texas wintergrass, tobosa, switchgrass, hairy tridens, tumblegrass, hooded windmill grass, maxmilian sunflower, silver bluestem, sand

2003 United States Production of 13 Crops

Crop	Prod. (Metric Tons)		
Barley	6,011,000	276,080,221	bu
Corn	259,273,000	10,207,022,425	bu
Dry edible beans	1,070,625	23,603,000	cwt
Cottonseed	5,847,000	2,890,296,200	lbs
Cotton	3,823,061	17,559,001	bales
Oats	2,100,000	144,676,875	bu
Peanuts	1,793,000	3,952,847,800	lbs
Rice	8,948,000	197,267,608	cwt
Rye	235,000	8,634,683	bu
Sugar	7,620,000	16,799,052,000	lbs
Sorghum	10,177,000	400,646,682	bu
Soybeans	67,179,000	2,468,380,390	bu
Wheat	63,590,000	2,336,508,567	bu

Total 437,666,686 Metric Tons

or

964,879,976,081 lbs

or

2,708,435,021 lbs/day

or

9.42 lbs/person/day

Cotton: lint basis;
Metric Ton 2204.6 lbs;
US Pop (est) 285,000,000;

Source: USDA, PS&D database; www.fas.usda.gov/psd
except dry edible beans, from NAS County Estimates; www.nass.usda.gov

Provided by NORM, 11/03

bluestem, buffalograss, curly-mesquite, sand dropseed, sideoats grama, blue grama, hairy grama, red grama, Texas grama, plains lovegrass, purple threeawn.

The wheat grasses, *Agropyron* species, are hardy, drought resistant and versatile, and capable of producing abundant forage. Almost all are perennials. These have great value in the northern Great Plains and intermountain region. This particular species dominates weeds, can be sown with little seedbed preparation and claims the pedigree nature. Old timers tell me that this grass reclaims abandoned land. This grass has about 150 species. The grasses that trap solar revenue and build meat protein with the help of water, air and a few earth minerals, fill volumes much as Wes Jackson's metaphorical DNA manuals would fill stadiums.

I can insert here a few that are explained in detail in *Grass: The Yearbook of Agriculture, 1948.*

Bentgrass, sometimes called redtop, takes its place among the forage grasses, redtop being a perennial. It matures in the same time frame as timothy. The bentgrasses grow all the way from Mexico to Canada, from the eastern seaboard to California depending on locality. This genre has other names — whitetop, fiorin, white bent, herd's grass, and others.

Meadow foxtail came to America about 1850. It looks a lot like timothy. It does not have much of a root system. It is one of the earliest cultivated grasses. Like most grasses, it takes on the potential of the area in which it is grown. Its production is best in late winter and early summer. In other words, it is a cool season grass, but also stands hot temperatures.

The beachgrasses, as the name suggests, made their deal with evolution on sandy acres, often near beaches. These grasses are tough perennials with creeping rhizomes. They grow on low nutrient soil, help control erosion, but keep cattle lean and hungry. In fact grazing animals should be kept away from such forage.

Bluestem is a household word because several species are excellent forage grasses. They grow almost everywhere as big bluestem and little bluestem and are often considered weed hay. The grazing quality diminishes as the plant matures.

The Broomsedge *Andropogon virginicus* species has a wide distribution, but unfortunately is a low fertility soil cover with poor palatability. It can be rated as a poverty grass.

Big bluestem is really a tall grass prairie plant, a Great Plains mainstay. Cattle grow fat on the leaves of this grass, and soils become securely anchored as moisture attends the sweep of the seasons. This prairie grass was routinely plowed under by pioneers, ultimately setting up a dusty wail in the 1930s.

Tall oatgrass is grown in the central and northern states. This is a hardy perennial, never propagated by root stalk. It grows best on sandy or gravelly land and more or less copies domestic oats, or vice versa. Experts say it has forage qualities, but it has never become popular. Cattle producers who need hay to bridge over snow seasons often consider this forage. Just the same, this one grazes well.

Carpetgrass is a coastal plains grass, and a resident of rainbelt states such as Arkansas. This is a perennial creeper. Stems root at each joint. Without underground stems it never becomes a pest in row crop situations. This one is valued for permanent pasture where moisture is abundant. It is popular wherever pasture can be intermingled with forests.

Grama grasses, not to be confused with gama grasses, exist as some 18 species in the U.S., all well established in the Great Plains, all good forage on range and pasture land. The grama grasses are summer growers, albeit subservient to moisture availability. Most classes of livestock thrive on grama grasses. The several species carry such names as Texas grama, Kansas grama, etc. Transfer of a North Dakota species to Texas and vice versa suffers the debilitation cattle suffer when transported from one end of the country to the next.

Sideoats grama is especially long lived and has a wide distribution on the Great Plains. It has a deep root system and uses moisture expertly. This one qualifies as "nature's benediction," to use Senator Ingalls' words.

Grass Economics

I would, if I could, have a tabulation on the economics of grass stated in terms of meat protein. The variables make what follows less exact (in terms of USDA figures) but the message is clear enough, just the same.

The seekers in the main have been content to scan the landscape for the biological diversity nature has evolved. Many of those mentioned below were catalogued by Ferdinand Roemer in the 1840s.

There was brome grass, for instance, 43 species, all native to the United States. Troublesome weeds are close and distant relatives. The name itself implies "a kind of oats." Most brome are highly palatable while actively growing. In Texas they call the grass Texas brome. Other popular names include rescue grass, an Argentine import of 150 years ago. This one is a short-lived perennial that lives well in humid regions. Smooth brome claims recognition as a sod grass with creeping rhizomes. It can and does live in Siberia, China, climes from which it was imported in 1884, and grows widely throughout the U.S. It endures low to moderate rainfall and low to moderate temperature. Bromes in general form a dense sod which resists erosion and handles cow traffic quite well. It also defies those two engines of erosion, wind and water.

Smooth brome does best on moist well-drained loam soils with high fertility. As with many grasses, brome is environment sensitive. Over the centuries, the survivors have staked their claims where you find them.

Wheat Grasses

Wheatgrasses sometimes confuse graziers because wheatgrasses are not the same as "wheat as grass." The wheatgrasses are drought resistant and versatile denizens of the mountain valleys and plains. Most ranchers consider the native wheatgrasses wonderful forage crops because almost all are soil building perennials often with running root stalks. The stems grow erect, resembling the posture of wheat. Many wheatgrasses produce sod and crowd out weeds, especially when seeded to establish new pasture. This one is so versatile — possibly because of the 30 species found in North America — and rates front-burner status as a soil protector. It can be considered a spring grass because of the lush growth that appears once winter weather departs. World-wide, there are about 150 species of wheatgrass.

Quackgrass is a wheatgrass. It is also considered a weed. It assails the psyche of farmers the way the dandelion troubles the urban lawn grower. Withal, *Agropyron repens*, or quackgrass, is often put to use as a soil holder in conservation practices.

Some species are annuals for which no use has been found. Many grass species will hybridize with common and durum wheat. The names are descriptive: crested wheatgrass, bunch grass, handles the cool, dry weather of the northern Great Plains and mountain valleys quite well. The grazing potential is great as is the root holding capacity. Delivering succulent feed when it is most needed is this plant's strong point. Crested wheatgrass can be a seed crop weighing about 90 percent pure with germination of 88 percent and weighs about 22 pounds.

Many of the wheatgrasses were introduced from parts of the old Soviet Union as a partial import of the world's 150 species. The full compendium would fill a sizable book.

Western wheatgrass is a native perennial, but its biography stays close to its botanical genera. It will grow on alkaline soils, but really prefers shallow, pelagic-type lake beds. Rhizomes take the root system to any source of water, soil fertility always governing the stand, its grazing quality and capacity for erosion control.

Classes of livestock like this grass, except during a dry part of late summer.

Bluebunch wheatgrass is also a native perennial. It has adapted itself to the western U.S. as a climax crop in the Northwest and the mountains of the West. It has been estimated that 60 percent of the green cover in some western areas is bluebunch wheatgrass. It is a drought resistant range forge that answers to the principles of holistic resource management. Beardless bluebunch wheatgrass is closely related, although lacking in spikelets. These two grasses are propagated only by seed, hence the necessity of livestock to stomp seeds into the soil. Success has attended reclamation of eroded or abandoned soil with these species.

There is another wheatgrass, slender wheatgrass, a native that achieved commercial importance. All classes of livestock relish this nutritious grass, on the plains, in the mountains, wherever it is native. Just the same, it is short lived and best utilized as a mixture.

In the meantime the business of perennializing wheat and sorghum and sunflowers and even corn proceeds. The Land Institute has opened a nursery in Argentina so the seeds can be harvested in the fall when it is spring here in order to speed up the breeding. Wheat will cross with about 100 different taxonomic entities. The diploid sorghum cross with Johnson grass, back crossed in the direction of sorghum has delivered the winter hardiness sought. The perennial sunflower has been crossed with the annual, although a stabilized population has still to be achieved.

Halfway across Kansas, the Arkansas River is dry. The sewage from Great Bend is now the headwaters of the Arkansas. Some 19 graduate students around the country are keeping Land Institute research on track.

There are so many variables in forage production based on genera and species rainfall, general environment and nutrient loads in the hundreds of soil types, computation of the solar energy harvest becomes next to impossible.

In *Unforgiven: The American Economic System Sold for Debt and War*, I made the case for grass as the key to prosperity unlimited. Grass trapped more sunshine than row crops accounted for. It nourished a cattle herd that demanded little more than sunshine, water and air, with grass responding to a few earth minerals.

Rhodes grass takes it name from pre-Zimbabwe Rhodesia. It came to the U.S. in 1902. It is propagated by seeds and stolons. It stands winter well when winters are mild, all along the Gulf coast, also in southern California and Arizona.

Bermuda, really a native of India, is now a resident of all tropical and sub-tropical parts of the world. Natives call it wiregrass, dog's tooth grass and devil grass. It's a long-lived perennial that propagates by runners as well as seeds. Bermuda grows on any soil if a modicum of fertility is present.

Orchardgrass came from Europe in 1716. As with most grasses, local names apply, oxfoot for instance. A perennial, it lives a long time in bunches. It seems to work in tandem with legumes — ladino clover, especially. It produces no stolons or underground rhizomous habitat in the entire United States.

The several species of wild rye call the Western states home. Most are perennial bunch grasses with a good sod-forming capability. Very palatable, these grasses are susceptible to drought. The wild ryes are known as Canada wild rye, giant wild rye, blue wild rye. The list works well in words, and all wild ryes are grazed best when tender and young. Each of the species define its preferred growing terrains, always with pH, nutrients, type of soil, the governing factor. Shade is often a factor as is the mix with other grasses and forbs.

Lovegrasses are reputed to encompass some 250 species and are ubiquitous as to habitat. They are generally not of agricultural value, exceptions being few. Weed manuals claim some species as plants out of place, a designation nature would hoot at. One might call them poverty grasses. They survive on soils of low fertility. Of the 40 species in North America, only about three are of any value, namely *E. obtusiflora* (New Mexico, the Southwest), plains lovegrass and weeping lovegrass. Weeping lovegrass came to the United States from Tanganyika around 1927. Another importation took place in 1934 for the Southwest.

Sand lovegrass is mostly a conservation soil building grass. Just the same it is palatable and is often overgrazed. This is strictly a sandy soil grass.

Centipede grass now has a toehold in the Southern states out to the West Coast. Spread by stolons, it is something between carpetgrass and Bermuda. Its moisture requirement to get started is high. It does not thrive in the North. Not much nutrition in this one.

There are about four fescues that invite attention. There are over a hundred species, according to USDA, all in temperate or cool zones. Some are annuals, some perennials. Perennials are excellent for forage and turf. Meadow fescue and tall fescue are superb. Red Chewings fescue and sheep fescue are very useful. Tall fescue flowers in June and July and likes damp pastures. It is distinguished from meadow fescue chiefly by height. Not a native, it was introduced from Europe. Still the nuances of fescue management can and have filled books and are so location specific they excuse further comment in these paragraphs.

Panic grasses fall under the genus *Panicum.* USDA references tell us there are 500 species, chiefly in warm regions of the world. Therefore the U.S. species are in the Southeast and the warm parts of the West. They are millets. *Panicum obtusum* is a long-lived perennial, survives drought but performs well where there are occasional floods. Livestock relish the plant. As with many grasses, livestock relish it when young and tender.

Buffalograss is an original, the grass Roemer or Lewis and Clark encountered in the early and mid-19th century. It is an excellent sod-forming perennial. As a species it dominated the upland shortgrass regions of the West, especially the central Great Plains. The buffalo thrived on the grass, maintained it in the manner described by Allan Savory. It spreads by runners and probably has a common DNA for much of an acre.

Indian ricegrass is a very tough native bunch grass, a resident of most of the western states. Once upon a time the Indians used its seeds for bread making when the corn crop failed.

Canarygrasses are both annuals and perennials. Southern varieties are annuals. The northern latitudes have summer annuals. Reed canarygrass belongs to wet lands. Herding grass, not a native and introduced as a forage grass, performs well where winters are mild.

Paspalum grass has about 400 species, but cannot be said to have much economic importance. Bahiagrass, dallisgrass and Vasey's grass often find their way into the forage mix.

Pearl millet and napier grass came to the U.S. from the West Indies. In the islands it was an important forage plant. Both are grazing and silage entries in parts of the South.

Timothy, once called "herd grass" when found along the Piscataqua River in New Hampshire, appears in the correspondence of Benjamin Franklin. This colorful grass probably got its name from Timothy Hansen, the man who transported the grass from England. By 1807 it was the most important hay grass in the United States. Its popularity has not diminished, especially in the northeast and north central part of the country. It also grows and survives well on the Rockies and Northwest. Nitrogen fat and protein decreases as the season progresses and the grass becomes less digestible and crude fiber increases with maturity. The best harvest is in early bloom.

Bluegrass has about 200 species, 65 of which are native to the U.S. Some are cultivated. It is, of course, a prime pasture grass, the subject of the paean presented in the fore part of this book. Many cowmen consider it the foundation grass for the paddock, its attributes so well known they hardly need further exposition.

Foxtail millet came to the U.S. from China by way of Europe. It became an important crop in the central states by 1849. Row crop farmers will recognize it as a close relative of foxtail weed, even pigeon grass. Depending on the locality, foxtail millet is called common, German, Hungarian and Siberian. German and common are produced on purpose. Others simply obey the biblical injunction to increase and multiply, often finding space in the pasture forage mix. Feeding value is best at first bloom, then fades quickly. As a hay, foxtail millet is ranked well below timothy. The USDA says "equal to prairie hay, but inferior to alfalfa and clover."

Sorghums and congeners, the Sorghum genus, rate attention as a forage crop. Sorghum, sudangrass and Johnsongrass are considered among the most important forage resources. These summer annuals tolerate hot, dry weather. Johnsongrass is a rapidly spreading perennial that holds up well where winter temperatures are mild. Most farmers consider it a weed in row crop situations, but it rates recognition as a pasture grass, especially in the South. Grain and forage sorghums are pretty much the same. Just the same, those with the most sugars are feedstuff preferred.

Selection and breeding have enlarged the number of varieties available on the market. Some sorghums are grown for industrial uses such as wall board.

Green sorghum feeding — either as pasture or silage — runs a risk of prussic acid poisoning, usually pasturing after stunted growth, then rain imposed the greatest risk.

Dropseed is a real native abundant in the southern Great Plains. The perennial species are palatable to animals. Consider this one a bunch grass. Its nutrient value is governed by soil fertility or lack thereof.

St. Augustine grass as it grows along sandy shores is merely another of the countless species nature has installed to hold the sandy shores in place. Many people use it for lawns.

The Stipa genus designates a needlegrass, habitat the central zone. Some 30 species grow in the West. This one ranks high as a forage grass on western ranges. Cows like the nutrients, but not the needles.

Zoysia is both the common and genus name for three species — manila grass, Japanese lawn grass and Mascarene grass. All are of tropical Asian origin. These grasses are generally used for lawn or turf purposes.

Supplements

Phil Wheeler, a long-time Michigan consultant, makes the point that most pastures fall short in delivering the balance animals require. His remedy for many pasture shortfalls is garlic — garlic in the paddock, garlic combined with yucca schidigera, a desert plant.

A second help for animals is parasite control with diatomaceous earth. It has already been pointed out that cobalt is almost always missing in U.S. pastures. Manganese in excess often inhibits the uptake of copper. Zinc out of balance completes the quartet of problems that permit brucellosis. In *Reproduction & Animal Health*, I have detailed the many known and some of the unknown roles of trace nutrients, 90 to 92 on the Mendeleyev chart. It would be a rare thing to find even 25 percent of the roster in the average pasture. Some of the chapters that follow will deal with the shortfall, the consequences, even the remedy.

Safe Pastures

Safe pastures are those that harbor very low levels of infected larvae. Safe pastures are ideal for young cattle. The usual way to create a safe pasture is to remove a hay crop during the first part of the grazing season. Second, grazing the pasture by a species such as sheep or horses the previous year or during the first part of the season.

And, of course, there is the ancient system of resting a pasture for one year. Relative safe pastures are often those grazed by dry cows. The mid-summer rise occurs during June. Moving heifers to safe pastures before this period will prevent exposure to large numbers of infective larvae. Heifers treated with diatomaceous earth before mid-summer are not likely to preserve pastures safely through the rest of the season. Usually one deworming is enough for a year. Mixed grazing can be practiced if sheep or horses share.

Grazing different species of animals together reduces parasite problems. Sheep and horses will graze close to manure pads where larvae concentrations are the greatest.

Cattle will graze close to sheep and horse manure, ingesting and killing their parasites. Alteration of annual species each year also suggests itself, switching species before the mid-summer rise is also effective. An additional word on controlling parasites is indicated. Experiments have revealed that there are many benefits to effectively controlling parasite in growing heifers. Research reported in early issues of *Acres U.S.A.* told of 30 to 60 pound gains over controls in split pastures. This is important because such heifers have an increased growth rate and reach breeding age earlier, and this results in extensive savings.

Bill Kruesi, author of *The Sheep Raiser's Manual*, once delivered a crystal-clear report on reclaiming pastures savaged by weed and bramble takeover. His recommendation was to use sheep, goats and horses as well as cattle to selectively remove noxious weeds. Goats graze best at shoulder height. This sentiment is akin to one expressed by Ann Clark in the section of this book that carries her responses to questions.

Kruesi wrote of frost seeding, that is reseeding a pasture when an evening brings on frost, and the day's sunshine briefly takes the frost away. Seeds settle in crevices and cracks accompanied by moisture. This seeding does not work well on snow because seeds become food for birds when not taken in by turf and soil and given room and board. Rotational grazing does the rest.

Two Recreational Drugs

California and several other states have passed laws that legalize the medicinal use of marijuana. These laws have been preempted by the Justice Department, and operators of state clinics have added their names to the greatest per capita jail population on planet Earth.

The reason for being of a general rejection of harsh sentences is grounded in the fact that this drug can deal with the nausea that goes with cancer chemotherapy, stop glaucoma blindness, ward off asthma attacks, and alleviate chronic pain.

Proscribed and jailed, growers draw heavy sentences for what the cow has known for centuries.

I can only speculate at the size of nature's manual for the DNA of this plant with its inventory of pharmaceuticals. With 400 chemicals it can assist a metabolic condition defined in terms of single-factor analysis.

Whether cannabis is a gateway to the use of stronger drugs continues to be debated, though research tends to show that it is not. Because of an atmosphere of intimidation, most seekers avoid the research just as they ignore the findings of the finest nutritionists the world has ever seen, *Bos taurus* and *Bos indicus*, the two genera and species of cows on the planet.

Laws and folklore notwithstanding, *Cannabis sativa* appears to be less addictive than heroin, cocaine, nicotine, alcohol or caffeine.

As a recreational drug, the common ditch weed doesn't rate much space compared to Viagra or the other drugs mentioned above. As a well-bred version produced as a companion plant in the cornfield or the clandestine greenhouse, it has become a cash crop rivaling pornography at the bottom line.

Like whiskey, the marijuana plant has its long-term effect, and this may separate the animal's use from that of the human being.

Delta-9 THC has been assigned more folklore than Greek mythology. Just the same, the risk to the pulmonary system joins tobacco smoking as a devastating assault on the breathing apparatus, with cigarettes delivering four to five times more damage than the Mary Jane puffer sustains.

Too much attention to this one plant, not even a survivor in the well-grazed pasture, may seem out of place in dealing with the pasture enigma, but so is almost every grass and herb when isolated in a box for purposes other than mandatory diversity in the pasture.

Wes Jackson's analysis of the prairie, row crop erosion, and the prospects of maintaining grain production with soil-preserving biennials demand much more attention than the fascinating story of a weed that has sent the underground economy reeling.

Jackson's Land Institute needs a high-profile assist, one on a scale with the push for the recreational drug that received former Senator Robert Dole's pharmaceutical attention. After all, Viagra can kill its users — dramatically, if also pleasurable.

These passages on the most controversial plant of the day are more than an aside. Although not a pasture choice, hemp has been suggested as a green savior of the environment. In Thomas Jefferson's day its fibers were used for paper-making, and such use today could save millions of trees. Grown in waste areas along most interstates, the alcohol from such biomass might fuel the entire automobile fleet of the United States, according to some calculations.

Cannabis

Across most of the Midwest and West grows a ditch weed with the scientific name of *Cannabis sativa.* The wild unpampered species has little commercial value as far as aficionados of the reefer trade are concerned, but this does not cancel out helicopter surveillance or botanical SWAT teams with cutting tools and a weed-free mindset.

Once a denizen of the Asian steppes, cannabis now thrives on every continent except Antarctica, spreading like milkweed or thistle, probably the most assiduously cultivated plant and has drawn the attention of more specialty growers. Cows like to nibble on its flowers, unaware of the draconian sentences that go with being caught. Cannabis is the oldest fiber crop ever cultivated.

The classical purpose for this plant was not to anoint pastures, but to make rope, canvas, clothing and paper. The plant is dioecious, having male and female plants in equal proportion. Given a chance, the cow will consume the female and male. These buds secrete a sticky substance rich in cannabinoids full to the brim with some 60 compounds. In short, cannabis is a medicinal botanical, not unlike fully half the weeds and forbs listed in almost any botanical manual. Some of these are psychoactive, such as delta-9 tetrahydrocannabinol, THC.

Ancient compendia tell of areas assigned to cannabis about 28 centuries ago, when its ancient seekers mixed cannabis with wine as an anesthetic used during surgery. Its milk was found to relieve labor pains during delivery. Except for the institutional arrangements, cannabis might be hailed as a miracle drug, this according to Lester Grinspoon of Harvard, in his text *Marijuana, the Forbidden Medicine.*

Solar Revenue

It is the flow of energy that governs economics, not the stock exchange or the will of the Congress. In the fullness of time, it is this flow that trashes legislative intent and makes a fool of errant public policy.

It is the entry of energy into the living world that concerns us most. Its name is agriculture and its instrument is chlorophyll, the green coloring matter in plants.

This is the key to natural energy. It transforms solar energy the way a power pole transformer steps down highline voltage for the task at hand. Clearly, the full inventory of the energy used to operate the planet, a nation, a people, comes from the sun. As E.O. Wilson has stated, there isn't a square inch of the planet that does not support life forms. The ocean itself may have several times more units of life per cubic inch than any land area. All rely on sun energy.

The route of that energy through plants, animals into human bodies and industry is clear and traces back to the sun much as economic progress traces back to raw materials.

That energy can be stored and recovered need not detain us. Suffice it to say a human being has to be supplied with energy before he or she can drum a finger. Economic life chiefly has to do with how nature empowers man. The amount of energy and heat that is spent demands a useful end product. Whatever the proximate source, energy arrives through plant life even when stored from an earlier geological period.

It would be possible to follow the trail of energy into countless *cul-de-sacs*, but not without concluding that agriculture is the key industry of life. Science, manufacturing, education, all have merely assisted, never as prime movers.

Machines merely imitate life. Even rain is the product of solar energy evaporating and condensing water, bestowing flow on rivers, generating electricity, servicing factories and assisting agriculture in sustaining service industries.

Solar income via chlorophyll dominates our economic system. This means agriculture. To effect a shortfall for agriculture income is to cancel out solar income to the tune of 70 percent of national income. Construction of debt-based banking and exchanges does not eliminate the necessity for newly monetized solar income each annum.

Stored energy — as in oil and coal — is spent as heat, but solar energy transformed into food energy reflects efficiency that makes all works of man pale into insignificance.

Production of raw materials creates capital. Stored capital, much like stored energy, becomes exhausted if not replenished.

Debt is only an ephemeral substitute for earnings. That is the meaning and source of inflation. It can now be computed that $100 on deposit in the fractional reserve system permits creation of over $13,000 out of thin air. The failure of the money supply to march hand in hand with the monetization of raw materials asks for the miracle of compound interest to work, and the process of consuming the capitalization upon which the economy rests to survive.

The shortfall of income for the raw materials production sector, grass included, has prompted public policy to endorse the blunder called World Trade Organization, North American Free Trade Agreement, and Central America Free Trade Agreement. These on the premise that an economy can import cheap labor via cheap goods and sell high-tech products to the rest of the world.

As the numbers indicate, the trade imbalance is staggering the system. Over 14 percent of the total income is occupied with foreign trade, yet a mature sense of values tells us this can never exceed 5 percent if stability is to be retained.

Public and private debt is now growing exponentially, doubling every five to eight years. National debt threatens to double in even less time.

The elemental constant in the statistics presented here is that the malaise traces back to what is happening to agriculture and grass. The solar energy revenue has not slackened. Nor have most of the venues that receive and transform it. The culprit is the accounting system that allows institutional arrangements to pay agriculture for less than the value received.

The smallest unit of account is the county. It used to be the acre.

8

A Seeker in Academia

Native range was a valuable natural resource long before white settlers were present in Texas. An estimated one million Indians lived and roamed the present boundaries of the United States at the time settlement by Europeans started. These natives supported themselves primarily from the wild game available in the forests and prairies. Although small patches of corn and other roots and grains were grown by some Indian tribes, the bulk of their food and most of their clothing and shelter was furnished by wild animals and plants. It is estimated that the country originally supported 50 million animal units of the larger herbivorous animals. Buffalo, deer and antelope were the dominant species of these creatures. Numerous other species of small game and fowl supported themselves directly or indirectly from native vegetation. . . .

One of the more vivid descriptions of vegetation in south and central Texas at the time of settlement was written by Dr. Ferdinand Roemer, a German scientist who studied the physical and climatic aspects of the country as it might affect future development by German colonists. He traveled from Galveston to Houston, San Felipe, New Braunfels, Caldwell, Fredericksburg, San Saba, Llano, and San Antonio from 1845 to 1847.

— Henry Turney, *Texas Range and Pastures*

"Find those wise old men out there," the late William A. Albrecht told me, "and you'll receive the lessons the university has forgotten." One of those wise old men taught a university course in forage and pasture before most schools even got around to considering a course in rangeland management. Henry Turney was not a Ph.D., for which reason the usual university press denied him the publication of his codification of lectures, notes and illustrations. The axiom that truth can't be removed with a hatchet won out just the same when Mrs. Turney assembled and published *Texas Range and Pastures — The Natural Way*, preserving the pardonable pride Texans have in Texas even though the message for the grazier is universal. In this exchange Henry Turney tells some parts of his story explaining why he hasn't fed hay to pastured cattle since 1947. His biography is contained in the interview below. Farmers with cows on grass will find new insight in what Turney has to say.

Q. In your book, *Texas Range and Pastures*, you seem to suggest that these principles that you are talking about are more or less limited to Texas. But with the 98th parallel dividing the 30-inch rainfall area almost from Texas to Canada, wouldn't these same principles hold for pastures in Kansas, Nebraska and the Dakotas?

TURNEY. I see no reason why they wouldn't. Of course, the season would be shorter, but I see no reason why the principles wouldn't be adequate for any of those areas, with rainfall as the governing factor.

Q. Is it correct that every 15 miles you go west of that parallel rainfall decreases about an inch?

TURNEY. I have heard that, and I think I have read it. I couldn't vouch for it, but it is the old historical statement that we have always made.

Q. In your book you cite Roemer, the German who came over to Texas around the 1840s. He seemed to have inventoried all the various grasses in the state of Texas as he moved from the coastal plain back to the hill country and into the semi-arid or arid areas. The gist of it seems to be that you do much better relying on natural grasses than you do on what we call "plain pastures," where you try to plant grass.

TURNEY. Those pastures are planted with introduced plants in a lot of cases. I can give you a little bit of background as to why

I did harp on that subject. I was raised on a small farm with grassland in Texas, and I went on to Texas A&M University. As soon as I graduated from college, I was employed by the Soil Conservation Service and ended up being a sort of territorial range man. A few years ago, I decided to retire. I bought some land here in central Texas and got myself a farm and a ranch. My good wife followed me out here, and we raised three sons here on the place, and this is where I live now. I was always interested in grassland pastures.

Q. You raised some cattle — is that how you came to raise pasture on your own place?

TURNEY. My dad knew nothing of what we call "range science," he had never heard of it, but he was a good cow man. I was raised in the country and watched him growing cows. My father didn't know much about grass, but he knew how to raise cattle, and he knew that the cattle ate grass. He relied on the grass. That was my early training in range management.

Q. Did he discern that climate had a lot to do with what species grew?

TURNEY. He was not what you would call well-traveled, but he traveled enough, and he was observant enough that he saw the variety of grass growing along the roadsides that had not been changed by man. He saw the difference between the Lubbock, Texas, area back down to what I would call east Texas. He knew nothing about range science, but he was a good range man, and he let the natural grass grow.

Q. In other words, he followed the old dictum that "the best grass is what grows in the ditches."

TURNEY. That's about right.

Q. When you go out into the hill country, that old rangeland seems to be full of cedar trees and various shrubs and brushes nowadays. What has happened?

TURNEY. I worry about that more than anything else. I live in the central part of Texas around the little town of Dublin, about 80 miles southwest of the Fort Worth-Dallas area. In this part of the country we had an invasion of mesquites, but mesquites were adapted from here to the west throughout Texas. I didn't know very much about cedar until we woke up a few years ago and found that stuff spreading all over this country.

Q. Is this because the grazing has been too intense on the rangeland?

TURNEY. It didn't help any. I would say that one of the major reasons for the proliferation of cedar and other woody plants on pasture is overgrazing, and the other, of course, is the loss of natural fires that would have held back the cedar, as well.

Q. The old story is that the Indians fired the prairie to get rid of the brush from time to time, and, of course, natural fires occur because of lightning. But nowadays we want to put out the fires.

TURNEY. Well, we really don't have enough fuel left on our rangeland to start a good fire nowadays. We have abused our rangeland primarily through overgrazing and also by never giving the land any sort of rest period, which the land needs to regenerate grass.

Q. And no carpet of residue on the soil to hold in the moisture.

TURNEY. That is a part of the problem because people believed that anything that was above the ground was cattle feed. They did not realize that the leaves composted and made the food that grew the grass.

Q. What has been your experience in observing how people manage rangeland or have cattle as a low-maintenance type of animal?

TURNEY. I'm speculating now, but we seem to be a people of numbers. Maybe it is human nature. When rangemen around here visited, their main question was, "How many cattle do you run?" They seemed to think, and still do to a large degree, that the number of cattle you own is correlative to the amount of income you make. We know differently, of course. We know it depends on how fast the calves grow and how much you have to feed the cows to raise the calves. If you have to spend a lot of money to get your cows to market, then you are not going to make much of a profit. I do think that there is more of a change taking place. I enjoyed my stay in the Soil Conservation Service, but I purchased my little piece of land and decided to go it alone. There was a man in the service whom I admired a great deal — you may have heard of him, E.J. Dyksterhuis. He had forgotten more about grass and range management than was known by all the other employees of the service put together. I did everything I could to pick his brain, and he was a good scientist. He ended up teaching at A&M. When I decided to set out on my own, I only raised as many cows as I could feed with the grass in my pastures.

Q. In your book you pretty much cover most of the grasses in the state of Texas. Many of them, we notice, creep up into Oklahoma and Kansas.

TURNEY. Let me give you an example of one of our grasses that also grows in other parts of the country, Little Blue Stem. I would call it the base grass of the black-land belt, which is from the Fort Worth-Dallas area down to Waco. There is a very fertile belt of deep, dark land there that was mostly farmland. Originally it was just covered with Blue Stem grass. It's a tall growing grass, and the cattle loved it. But it was killed out on most rangeland in this part of the country by overgrazing. I imagine you have some Blue Stem up there.

Q. We have it in the Flint Hills in Kansas.

TURNEY. That's noted for the Blue Stem. We overgrazed the native Blue Stem here, and it's mostly gone now. An example of a grass that survives is Buffalo grass, it grows close to the ground and escapes grazing, but Little Blue Stem grows very high, and it didn't have a chance.

Q. The old saying in the range country, especially up in Montana, is "he's all hat and no cattle." Of course, in Dakota the story is that east of the Missouri River everybody wears a baseball cap, and west of the river they wear a cowboy hat. That goes with what you were saying about how many cows a person has. You are saying that you may have a more economic operation with less cows that are grazing natural grasses than you do with all the hot-shot systems that are being touted by academia.

TURNEY. I have been around a long time. I have used and, in some cases, even recommended some of the introduced grasses, but I don't know of any introduced grass that can compete with the natural grasses. That is my opinion, and I firmly believe that. In Texas we have what we call K.R. Blue Stem. I don't know where it was first developed, but the King Ranch started growing it. It was an introduced species, and the reason they started it was that it grew close to the ground and continued to grow in spite of grazing abuse. It is still one of the popular introduced species here, but it is nothing like our natural Little Blue Stem.

Q. Was this more or less a form of no-till that was introduced to keep that carpet on the ground?

TURNEY. In most cases it was planted on land that had been in cultivation. The owners wanted the land to convert back to grass, and they would seed it to K.R. Blue Stem.

Q. And that worked?

TURNEY. Yes, but as far as feeding livestock — and a lot of people will argue with me on this — it does not produce the type of feed, in volume or palatability, of our good old natural Little Blue Stem. This newer grass went all over Texas. It will flatten out and does protect itself to some extent from overgrazing, and our Little Blue Stem does not.

Q. In your own experience, what kind of a cost have you been able to achieve in finishing out cattle on grass?

TURNEY. I haven't had experience with that. My experience is with the mother cows and selling calves after the weaning stage. Depending on the amount of grass the mother had, the selling weight was from 400 to 600 pounds.

Q. In that situation, what is your cost going with the natural grass?

TURNEY. That varies. Quite a few people have so grazed their land that they have to go to feeding in the winter time. They have to buy hay and cottonseed cake, which is a high-protein abstract from cotton. If a man takes care of his grass, he does not need to bale hay. A lot of people in this country would laugh at that because they run out of grass in the wintertime. I have been out here on my own for the last 40 years, and I don't feed hay at any time. I guess I would, but we don't have as deep a snow here as they do in other places, and cows can go a day or two without grass. A lot of people have to hay their cattle even in the summertime because they run out of grass, and that is an expensive proposition.

Q. Is that because they failed to rotate their grass?

TURNEY. Not only did they fail to rotate their grazing land, they allowed the grass to be bit off as soon as it would stick its head out of the ground.

Q. They didn't ever allow it to come back? The buffalo had more sense than that, didn't they? The buffalo would graze an area and then not come back to that area for a year.

TURNEY. I do believe the modern farmer is getting smarter and learning some of these lessons. During drought periods a lot of cowmen down here went out of business because they did not have any surplus forage. They had no choice but to go out of business. It would have been a whole lot better if they had taken better care of their grass. I've been out here raising my own cattle and watching my own grass for the last 40 years, and I've never had a bale of hay out here. Even if it snows, it never lasts longer than two or three days. As I said, a cow can go that long without forage although they may need concentrates during that time.

Q. Understanding this, you have to first understand what rainfall area you are in, is that correct?

TURNEY. That's right. Our rainfall belt is pretty much north and south down here. The further west you go, the less rain you have.

Q. In the Kansas City area you have 36 inches of rainfall a year. As you move west past Topeka, you no longer have tumbleweeds coming east. If you get to western Kansas, you are back to 17 inches of rainfall. So that works right along with the pattern that you have moving across Texas. You are talking about annual rainfall, about seasonal distribution of rainfall and temperature, and humidity.

TURNEY. And there are variations of that. Around here we have wet years and dry years. Some years it just seems like we have rain every time we need it, and another year it will look as though it will never rain. You also have changes in soil and climates, but no matter where you are, cows eat grass.

Q. Nature pretty much invented no-till farming, because if you have the grass coming up without being overcropped by that cow, then it will overpower the brush trying to get a foothold.

TURNEY. It will certainly slow down the invasion of unwanted woody plants.

Q. Such as the cedar.

TURNEY. Yes. The natural fires also kept the cedar back, but like I said before, we just don't have enough fuel on the ground to fuel the fire because the ground is so overgrazed.

Q. Why did you write your book?

TURNEY. I was out here minding my own business, and the son of a good friend of mine, who was with Tarleton State University in Stephenville, Texas, came by my place and wanted to know if I would teach over at Tarleton. I said, no, I'm a rancher and a farmer. Well, he kept asking if I couldn't take a little time to teach a class, and finally I agreed to teach a course in range management. Lo and behold — and I get mad every time I think about this — there wasn't a decent text that I could find. All of those Ph.D.s who had been teachers all their lives at A&M and Texas Tech and those places, and they didn't have a book. I don't know how they taught. That is how I came to write the book.

Q. How long did you teach over there?

TURNEY. Twenty years.

Q. How come a Texas University press didn't publish your book?

TURNEY. My wife says it is because I didn't have a Ph.D.

Q. Well, doesn't that tell the whole story? If you want the real information, you find some old timers out there, some wise old cattlemen who know more than the academics do.

TURNEY. I think that's right. That is the reason that I wrote my book. I was teaching my course two afternoons a week, and I needed a text.

Q. What would be your advice to a cow/calf operator, no matter where he is?

TURNEY. My first advice would be to give parts of your grazing land a good rest during the summer months — or a good part of the summer months — at least once every three years. I try to rest mine once a year. That is just an old farmer-rancher's opinion right there. That is what I ended up doing, and it is one reason I have never had to feed hay. I have some natural grasses that have had no cattle on them for the last three or four months of the summer so the grass can grow. In the northern part of the state, what we call the Panhandle, the winter weather may require that they feed some hay, but not here in this part of Texas.

Q. Of course, if you move up into Nebraska and Dakota, you are going to have to run your equation differently.

TURNEY. Oh, you bet. It varies with the type of rangeland you've got.

Q. But it still doesn't depart from the idea of staying with the naturally evolved grasses. Isn't that your message?

TURNEY. That is right. Some people will argue with me on that, but I will stick to it.

Q. Sabino Cortez mentioned that you were able to bring your calves up and only have about $50 invested in an animal by the time you sold it. Is that still correct?

TURNEY. Yes, I think he was talking about feeding the mother. But yes, all the expenses, all the money that you are out adds up to about $50. That's right. Now, if you have a dry year, you might have to feed more cottonseed cake, but in good years you don't feed very much at all.

Q. But the bottom line is that you do not bother with balers and hay and mowing machines.

TURNEY. There hasn't been a bale of hay on this place, and I haven't had a cow starve to death, either!

Q. What is the best approach for someone getting into the business? Would it be to find out what the natural grasses are in the area?

TURNEY. Yes. There may be some areas where introduced species might be best — for instance, in an area that was never range or pasture but was forest — but not many. In east Texas I think they've gone too far with introduced species, but I can see why they had to use some grass seed. If you go from east Texas to west Texas, you do best to depend almost entirely on Mother Nature. When I would buy a plot of land to join what I already had, I would seed the fields to grass, but I used native grasses. Little Blue Stem was my base, and then I'd put some sideoats grama seed in there, and on the slower places I seeded Indian grass. The other thing is to keep the number of cows you own in line with the amount of grass you have to support those cows — it is more economical to have your land feeding your cows instead of you buying their food.

Q. This is the point. While you titled your book *Texas Range and Pastures*, the fact is that the principles that you are talking about are valid from top to bottom, from Canada to the Rio Grande.

TURNEY. I think so.

Q. You are one of the wise old men, and we need to know what your lessons are.

TURNEY. I will say this. The bookstore man at the university used to complain to me because very few of my students turned their books in to the bookstore for resale. He would jokingly tell me that I was making him go out of business. It was very flattering to me that my students wanted to keep their textbooks.

Q. And the bookstore guy wanted to buy them for nothing and resell them at nearly full price.

TURNEY. Yes, and he was complaining about that.

Q. Well, your little book is a good one.

TURNEY. I didn't have anything to do with that. My good wife took all my manuscripts and old mimeographs, and she's the one who published the book.

9

Economic Success for the Small Farm

In order to explore the thinking of those involved with academia, I uncorked several Socratic-type interviews, all of which appeared in *Acres U.S.A.*, the objective being to find how well practices in the field were connecting with advice handed off in the schoolroom. That the most progressive Ph.D. approach was right on target became evident in the answers harvested from Dr. Ann Clark.

Economic success for the small farmer lies in reducing the cost of production and increasing the profit margin, as well as getting higher prices for the product, whether it be crops or livestock. Teacher and specialist E. Ann Clark, Ph.D., believes that managed pasture is a great way to achieve both of these goals.

E. Ann Clark is an associate professor of plant agriculture at the University of Guelph, Ontario, Canada. An American, she has been a Canadian teacher for almost 25 years, and now specializes in pasture management. Her answers to my questions fill in the details. Most of them suggest a drastic change in the model now being followed by mainline farming.

Q. Dr. Clark, Archer Daniels Midland (ADM) is running a series of ads in which they're telling farmers to close down their pastures and turn them into soybeans, because they can grow more protein that way. What would your comment be?

CLARK. Well, I don't agree with that at all. First off, what's of interest to a farmer isn't growing more protein; it would be making more money, and I think the people that are really being taken to the cleaners are those producing raw materials, whether it's raw soybeans, or raw corn, or raw milk, or raw anything that then goes into a bulk pool. They're very much at the mercy of the buyer of that bulk pool such as ADM. So I think that's an inappropriate type of advice for a farmer to listen to if they're interested in staying in business.

Q. What is the alternative to this commodity approach to agriculture, then?

CLARK. Well, certainly in my own opinion — of course I am a pasture scientist, so my bias is for pasture, I'll say that up front — but I think a value-added product such as meat or milk is worth more than raw grain in the first place, and even more important than widening the margin by lowering costs — which is what pasture lets you do — it also gives you the opportunity to create a niche market, or a branded beef, or some kind of specialty product that lets the farmer retain a higher fraction of the value of the commodity, whatever it is. What is needed is to distinguish your product in some way, such as grass-fed beef, or organic beef, or Wellington County pork, or whatever it might be, but something that lets you sell to clients, and your own guarantee is on it — you or the cooperative that you're working with, or whatever it might be.

So, I think certainly the key to farmers is two-fold, if you want to stay in business: first, reducing the cost of production and increasing the margin, so that you're making more money for each unit that you do produce; and second, selling it to people that are willing to pay for it as a value-added product, something that's other than just a raw material. The problem that people don't seem to realize is that ADM and other buyers, when they take those products, they then transform them into value-added products, and they retain the lion's share of the value. In Canada right now farmers are retaining less than 10 percent of the farm-gate value of what they grow and what they're selling. The other 90 percent goes predominantly to input suppliers and the balance to the bank.

Q. In other words, agriculture is being organized to take care of the input suppliers rather than the primary producer.

CLARK. Well, that's exactly true at the production level, and then beyond that, you get into the retailers and marketers and packagers and all that — when you or I go to the store to buy a loaf of bread, I think that the value of the wheat is like 3 cents, and the rest of the dollar and some cents that you pay is in packaging and retailing. It's amazing how little is left for the farmers, and this is because of this bulk-commodity orientation that farmers have been told to patronize, which is really to their detriment.

Q. Economics aside, isn't there something inherently wrong with this near-total reliance on grain for feed-lot animals and almost a rejection of the idea that herbivores should be eating herbivorous crops?

CLARK. Well, I can't agree with you more, and not just herbivores. But certainly, one thing that people doesn't really realize is that between 70 and 90 percent of the grain that we grow in North America is grown to feed livestock. It varies with the grain type, but that's the sort of fraction, so a really large proportion of what we do in North America just feeds livestock. Globally the figure is 40 percent, so 40 percent of all the grain that's grown on the entire planet is fed to livestock.

It's a real travesty and a shame to me that so much of the livestock that are consuming this grain don't need to be consuming it at all; they're either herbivores, as you say, which would be things like cattle and horses and sheep and so on, or omnivores, like pigs, which can eat essentially all of their diet as forages, but they do better with some grain, and even chickens can displace a fraction of their grain with bugs and grass and clover, so it's really unhealthy for the animals to be on the sorts of diets that we've concocted, whether it's grain or by-products or whatever.

It's also ecologically unsound, if you think about all of the hazards that there are in agriculture that relate to growing of grain — those hazards are unnecessary, whether it's annually opening up the soil to erosion and degradation, or the pesticides that we're using, the energy costs of working the land and then planting the crop and harvesting the crop — all of those things could be greatly reduced or even eliminated if animals were out on grass, grazing.

Q. Wouldn't we eliminate a great deal of the pollution in the groundwater and rivers and streams?

CLARK. I think that that is a true statement, but I think it needs to be tempered a little bit, because although pasture has a lot of advantages, one area that I'm not convinced we're 100 percent safe on is nitrogen and leeching, particularly when you get to high-production systems with very high stocking rates and high nitrogen. Whether it's from clover or from fertilizer in a humid environment, there is some scientific evidence of leeching to groundwater from pastures, so I think this is an area that needs further investigation.

I think there are probably ways around it, but this is not something that we're doing very much research on — I know there is some work at Pennsylvania at the USDA Pasture Lab there; there's some work also in Minnesota on this, and more work needs to be done. But certainly I think it's fair to say wherever you look at the profile for nitrogen leeching to groundwater, it's always concentrated in zones of corn and, to a lesser extent, wheat, and certainly in confinement feeding systems.

Q. Supposing you have crops and you're not making anything on them; you're losing your shirt, in other words. How would you convert that back to pasture?

CLARK. Well, it's not that hard to do. Now if you had some background or experience I would repeat Alan Henning's advice — Alan Henning is a well-known grazing consultant who spent 12 or 15 years in New Zealand — and when farmers in the Midwest ask him how to convert ground to pasture, what should they plant, his response is, "Nothing at all," and I love that. It's not what the farmer wants to hear, but I'll just repeat the advice that he gives them, which is that the farmer may have some old pastureland or something adjacent to corn land, and Henning tells them to take the old pastureland and split it into halves, and one half you graze rotationally, and the other half you let go to seed. And what you do is, you graze for 3 or 4 hours on the good part of the pasture, and then another 2 or 3 hours in the seedy part of the pasture, and the rest of the day they go stand on the old cornfield, and they poop out the nutrients, and they poop out the seed, and over time the pasture establishes itself.

I have had some farmers say it takes too long to do that, and that may be true, but another example that I'll mention that's sort of along the same lines is an excellent grazier in New York state. I've heard him speak twice, and his income is based on abandoned land — land that is going back to trees in New York state. He goes and finds the owner, gets permission to fence it, and puts up a good five-strand high-tensile wire fence, and then takes a brush bush-hog and knocks down anything that can be knocked down, cross fences, and starts rotational grazing, with no seed, no fertilizer, no nothing.

Q. Using the New Zealand plan?

CLARK. Pretty much, except he doesn't feed it, and it's just whatever is there naturally, and he custom grazes, so he doesn't own the land and he doesn't own the cattle, and he makes good money doing that.

Q. But the cattle would be distributing the seeds in the way you previously described, wouldn't they?

CLARK. That's right, and of course the other trick that people don't recognize is the dormant seed bank in the soil — there's just a phenomenal amount of grass seed and weeds of all kinds in the soil. The seed is there, and all you have to do is remove the trees, and they start growing. So if you've got enough of them, and with good grazing management, you can generate a really nice sod in a short period of time. Then he's laughing — you know, I mean he's doing really well with it, and I admire what he's doing.

But in practical terms for someone that wants to convert, say, grainland into pasture, I personally would do it the cheapest and simplest way possible: broadcast seeding, trampling it in with livestock. If they have no experience at all with grazing, a very simple rotation of two or three or four paddocks is plenty to get started. They should do their homework, do some reading, talk to some people, find some neighbors that can advise them as far as how to get the fence up, and where to put the gates, and where to put the water, and things like that — because those things make a huge difference on the ease of operation of the thing — and then get started.

Start real simple, and make your own mistakes, and put out as little money as you can. Where I put my money is in the perimeter fences; the interior fences can be very simple, one or maybe two strands — a temporary fence — and half the time I never make it to permanent fence for the interior, because the temporary works so well, and it's so cheap.

Q. What kind of a mix would you use in trying to reseed a pasture? Let's say you decided you are going to do some seeding other than livestock?

CLARK. Well, certainly, the first point to note is that what I would recommend here in Ontario is not necessarily what would work in southern Wisconsin or Illinois . . .

Q. Or Kansas.

CLARK. So take it with a grain of salt, but in general, I like more complex mixtures, five or six species, carefully chosen so that their maturity is matched, and their palatability is matched. I always use one or two legumes — always white clover where we are here, but if you're in a hotter, drier place it might be red clover — because some kind of legume is critical. I don't have good luck with keeping alfalfa and trefoil in the mixture, but that's because I'm intensive rotational grazing, and I think those species need a somewhat different management.

Another thing I always put into a mixture is a bottom grass — Kentucky Bluegrass works very well for us here; if you were in a wetter place, perhaps perennial ryegrass — but not very much, two or three pounds is plenty, and it establishes very fast, so you'll get your bottom very quickly. I think an aspect that people don't put enough emphasis on is the importance of a good sod to hold up the animal's weight, particularly under wet conditions.

Q. But you would be able to get the natural nitrogen cycle working under that system, wouldn't you?

CLARK. Yes, particularly with legumes, but you'll certainly need to bring in some legumes if the land has been out of pasture for 10 or 20 years. The nice thing with legumes is the seed has a very protracted dormancy interval — it's measured in decades. So not infrequently the first thing people notice when they start grazing is all of a sudden they've got clover that they never had before and they never planted — it's the dormant seed that germinated, and it's growing. But the seed is relatively cheap, and you know you need a pound or two of it, and it's real nice to have it in there and fixing for you. And then, where we are here, it's a continental climate with some Great Lakes effect, so we need a couple or three taller, coarser grasses that are more drought tolerant, things that will perform a bit for us in spring and fall, but certainly in the summer.

If the grazier has some experience, I would recommend orchard grass; if they have no experience, I would not recommend it. It's a hard species to manage well, but it really returns tenfold over in animal performance for someone that knows how to manage it well. Each species has its strengths and weaknesses, and no single mixture of one or two species is going to do what you need.

The other thing that I recommend to producers to think about is somewhat different mixtures for different parts of the farm. Every farm has highland and lowland and north-facing and south-facing and near the barn and away from the barn. Different parts of the farm may perform better for you in spring and fall than in summer, so those can be designated as your spring and fall pastures. Other places, like lowland that's too wet to get on in the spring and is still growing in the summer when the highland is high and dry — that can be your summer pasture. That doesn't mean you're only going to use it in the summer, but its main strength would be in the summer, so you'd put in a mixture of deep-rooting species, things that are not going to be strong in the spring, like bird's-foot trefoil as an example, and that you can count on for getting your summer grazing there.

So, I think different mixtures on different parts of the farm — complex mixtures relative to what many extensionists would recommend, and always one or two legumes, always a bottom species, and then other species as fit to that particular location.

Q. How do you compute carrying capacity?

CLARK. Well, it depends on how much you know about the place. I do it on a percent of body weight basis, so if I've got a 1,200-pound cow, and let's say she's lactating with an actively growing calf, she might eat 3 percent, 3½ percent of her body weight per head per day in dry matter, so I would multiply that times her body weight and say that's the amount she needs to eat, and I add on 10 or 20 percent wasteage from trampling and residue and so on, and then multiply the number of animals and say, if I've got 2,000 pounds to the acre of herbage on offer — that's to say, herbage above the residual height that I've designated, you know, the lowest that I'm going to graze; I don't care how much is the total there, I want to know how much is on offer within the grazed horizon — then you can calculate how many days, how many animals can stay on that piece of land. And that's the way I do it.

Q. Have you done any work that puts a comparison to managed pasture as opposed to row crop in terms of return to the farmer?

CLARK. Yes, actually, that was one of the first things I did here at Guelph. We did a three-year stocker trial that was on 50 acres of land, and we actually had several different treatments, but on average we compared about 300 stockers altogether over the three-year trial, and I costed it out using a full enterprise budget, including the land, and the interest on the cattle, and the negative selling margin, and the whole thing, and then compared that to the provincial averages for grain and all the different grain crops, and hay, and so on, and we made quite a bit more money. In fact, some of them were losing money on average, and over the three years we netted about $75 or $80 an acre. Some years it was higher than that, and one year we had a 30-cent negative margin, so we lost money that year, but it averaged about $70 or $80 an acre.

Q. This is not computing anything for minerals or veterinary bills?

CLARK. No, we included everything, even the return to management was included in that. I had everything lined out to yardage fees and trucking fees, and you know, fertilizer. I amortized the fencing over 3 years. It was a very rigid budget.

Q. What was your experience, though, with veterinary bills?

CLARK. Well, we didn't have much. Oh, in those days we were implanting still, so we had implants on the steers, we had salt and mineral out, occasionally you'd get pink-eye, very occasionally hoof-rot, but really nothing. I think we lost one animal to lightning, and these were yearling steers, Hereford cross steers, but they came to us healthy, and they did well. Their starting weight was about 650 pounds.

Q. Have you done any research on dairy animals in connection with your pasture approach?

CLARK. Not much. I've had considerable difficulty getting funding to do that. In fact, I lost a considerable amount of money in research funding through the province here because I proposed to do it to dairy, and the dairy people succeeded in getting the money taken away from me, because they argued there was no place for pasture on dairies today, and gave it to corn, so I lost about three-quarters of my research money.

Q. But isn't this a core problem with animal production or animal husbandry today? This business of trying to grow them in confinement, and they're manufacturing strains of *E. coli* we haven't even counted yet, and things like that?

CLARK. Truthfully, I don't know a great deal about confinement, but the problem that I see is that the people that are in power are people who have a vested interest in the status quo — in keeping things as they are and continuing on the trajectory that they themselves have invested in. And so someone who's already happy with what they've got and with their brand-new harvester, or silo, or their brand-new liquid manure system, or whatever it might be — they're going to resist any movement in another direction, and that's what happens.

As far as animal health goes, I think this is a real time bomb — the prophylactic use of antibiotics, the high-stress conditions these animals are under, the lack of exercise, the feet and leg problems, and so on. I went out to a major dairy conference once to give an invited talk on pasture — I won't say where it was, because I don't want to get anybody mad at me, but it was very large and well attended, and a lot of knowledgeable people there — but there were about 30 speakers invited, and I was the only one on pasture, I was kind of a token pasture person. So I gave my talk to about 500 very stony-faced dairymen who didn't want to hear what I said, and then I had to sit through several more of these talks, and it was astounding to me — the talks were all about taking acyclic cows and turning them into cyclic cows with different kinds of injections of this and that, really extreme measures.

As a mother myself, I'm listening to all this and thinking, "These are all men! They've never given birth. They don't know that when that cow doesn't cycle it means something — she's under stress. And rather than just back off on the stress and reduce the conflict between lactation and gestation, because that's what the problem was, they were trying to force her to both lactate *and* cycle, and it was astounding to me, the extremes that these fellows are going to to get these poor cows to cycle.

It was really pushing rope, and fortunately for me there were about 20 or 30 grass people in the audience, and we sort of escaped the conference and went out and looked at some grass while I was there. But I think it was a real good indication of the way things are. I mean, if you go to any conference where it's confinement, there's a lot of gloom and doom, and there's a lot of sad faces, and a lot of stress. And when you go to a grazing conference, it's like a breath of fresh air — everybody's excited, they're trading stories, "This worked for me," and, "Why don't you try that?" — you know, it's wonderful.

Q. In your talk before the 20th annual meeting of the Ontario Organic Group, you seemed to call into question these practices that seem to be anti-grazing, and you mentioned some consequences that extended well beyond agriculture into civilization itself. I wonder if you would recap those.

CLARK. Sure. One of the tragedies that has occurred recently in Ontario was a water pollution incident in Walkerton, which is in western Ontario. So Walkerton had experienced a situation of coliform contamination, and it was the pathogenic *E. coli*, as it turned out, in their water supply. Seven people died, and many people became sick, and some small children have permanent problems now because of it.

It turned out that the pathogens were traced back to a particular farm, and it was in fact a pasture farm that this happened on, and it was a veterinarian that was the owner and the operator. So what I said in the talk was, the problem was not the cattle, and it certainly wasn't this poor operator, who was getting a great deal of negative press at the time, but rather, the high-grain diet the cattle are on — not necessarily on his farm, but wherever he got those cows from. The practice of high-grain feeding has been associated with an environment in the rumen that favors the development of this pathogenic *E. coli*, and I cited some research from Cornell University on this issue.

There was quite a bit of hoopla about this in the press afterwards, and in fact some of my colleagues in pathology in the vet school here disputed what I had said, saying that, yes, it is true, that's what the Cornell guys said, but there's other research that suggests not, and it's not a done deal, and I accept that — I think that's a fair point. But I do think, and in fact more work has come out of Cornell recently suggesting that a high-grain diet does create conditions in the rumen and in the gut which are favorable to problems, and that animals that are on forage-based diets have a more normal rumen pH and a more normal rumen environment, and they do not develop these problems.

So I think from an animal health point of view, and from a human nutrition point of view — because we eat the meat and the milk from these animals, let alone the CLAs, and nutrition, and so on, the benefits that come from it — it all points in the direction of pasture. So then the question is: Well, if it's that obvious, and if it's that critical, why in the world are people resisting pasture? And I think the answer to that is going back to your opening comment about ADM. I think that large companies — and I don't single out ADM on this, I think there are a number of players, but they're one example — have found it convenient to drive wedges that have basically isolated conventional farmers from everywhere else.

The wedges go between conventional and organic, and between farming and environment, and between rural and urban, and by the time you drive all those wedges, you wind up with conventional farmers sectioned off into a little corner or a little cell where they can only trust people like ADM and Monsanto and Pioneer, and so on. Those companies portray themselves as the farmer's friend, but in point of fact, the opposite is really true. It's not to their benefit to see agriculture healthy and thriving; it *is* to their benefit to see agriculture dependent on them for input and for markets, and it's to the benefit of the large companies. That's really where the money in agriculture is today, and there's quite a bit of strong evidence to support that.

Q. That's true, the suppliers seem to be increasing exponentially.

CLARK. Yes, John Ikerd and others have shown that in fact most of the money that's in agriculture today is in the hands of input suppliers and retailers and marketers, and it's just a tiny fraction that remains in the hands of farmers. So farmers are not really being well served, certainly by industry, and certainly not by government. Somebody once said — I thought this was so brilliant — that agricultural researchers must be the only class of professional that rates its success on how many of its own clients it can drive out of business.

Q. Well, the intellectual advisors of agriculture seem to have that attitude, and yet, as you pointed out in your Ontario talk, there's getting to be a great deal of rapport with the consumer and the organic wing of agriculture.

CLARK. Yes, I think that's sort of a natural niche, I mean they come in together, and some farmers, as well, are moving in that same direction, not necessarily for the same reason, but I do see a kind of a coming together of people who have felt disenfranchised by mainstream agriculture in one form or another, either because they can't make money at it anymore, or because they're afraid of things that are in the food that we're producing through mainstream agriculture. And certainly you get the baby-boomer effect, and people with money and willing to buy and so on, but I think it's a reasonable point to be made that organic agriculture has to be accessible to everyone, not just to the well-heeled. So, I think we need to come up with ways to make sure that there's enough organic food for people who want to buy it, and that they can get it at a price that is reasonable.

Q. Why are the institutional arrangements for providing information so tardy in recognizing what you're saying?

CLARK. I was just talking to a group of Dutch consultants on this point. I guess there are several things. Universities have become quite collegial with industry, basically, and this is not an accident. This is not to speak ill of universities — I am a university professor, and I've invested 25 years of my life doing it, so I'm one of 'em! — but governments have intentionally withdrawn government funding in the public interest and mandated that in order for those of us in universities and government to gain access to what little money remains in the government purse, we have to get matching funds, and that means we have to find an industry sponsor or a rich philanthropist — usually it's an industry sponsor — to underwrite part of the bill, and it's usually 50-50 or 75-25, or whatever it might be. So that means if you're going to get an industry to pay for the research, it effectively channels public researchers to the service of industry, and being in the service of industry can — although not always — be in fact a disservice, be in direct contradiction to our mandate, which is to serve the public.

Q. Or find a body of knowledge that's suitable to that particular interest.

CLARK. Well, that's right. And this is not to say that the information is invalid or that they're lying or that there's any malfeasance involved. It's just that the questions that you can ask are determined by who's paying the bill. Genetically modified organisms (GMOs) provide a good example: it's virtually impossible to find money to do research on the environmental risks or food safety risks of GMOs, but it's easy to get funding to do research into potential applications of GMOs. This is not to say there are no risks; it's just that you can't get the money to do the studies, because the companies that own the GMOS, it's not to their advantage for those risks to be identified.

I'll just give one quick example: you may have heard of the case of Arpad Pusztai. He and his wife, Susan Bardocz, are eminent nutritionists who were employed at the Rowett Research Center in Scotland, and he got funding several years ago to conduct a very large trial involving rats being fed a particular kind of genetically modified potato called snowdrop lectin potatoes. The long and the short of it was that he became alarmed at what he saw in his research and spoke up about it. He did this with the permission of his director, and he did it twice. The second time, he was fired. His wife, who was actually his boss at the Rowett, also left at about the same time.

Now, following on from that, it's been very interesting to see. There have been no refereed publications, no scientific peer-reviewed publications, that have followed on to examine the work that he did. Because he's been strongly vilified and portrayed as a bumbler, as devising a stupid experiment — "of course rats don't like potatoes," and all this stuff — really cruel treatment for a man of his stature, because he's a very eminent nutritionist. Getting back to the story, Susan Bardocz went to the government of Norway and asked for funding to repeat the trials and to do it properly and with male and female rats, do it for five successive generations, you know, really a first-rate trial. She got the money, got the lab, got everything set up, and the whole thing is hanging fire, because not one company will give her the seed that she has to have in order to do the trial.

Q. In other words, we've arrived at the situation: "This is the case on which I base my facts."

CLARK. It's a very alarming situation because it may very well be that there's nothing wrong with GM foods — what do I know? I'm not a nutritionist.

Q. We don't know, though, do we?

CLARK. But the fact that they won't give her the material that she needs to even ask the question in an open and above-board fashion fills me with concern. And that is one example of "he who pays the money calls the shots," and this is what happens when governments oblige researchers to get matching funds in order to do research. As a researcher, then, your choice is either don't do any research and become 100 percent teaching, or total lying and doing what industry says, or, as I've done myself, you learn to ask different questions, questions that are cheaper to answer, things that you can do with the resources that you have. In my own personal case I've also opened up a consulting company, and the proceeds from it fund part of my research. So that's the way I've gone about it, but that means of course that the sorts of questions that I ask don't require expensive equipment and dozens of technicians and fancy labs and that kind of stuff, because that funding is simply not there.

Q. In the meantime, the farmers out there are proceeding on a pragmatic approach, finding answers on their own and sometimes even codifying them, is that not correct?

CLARK. That is correct, especially in the area of organics. At that talk I gave at the Organic Conference at Guelph, I was praising the fact that the organic industry is a very resilient self-reliant group of individuals, usually operating as individuals, not as a cohesive whole but as individuals who do their own research, do their own teaching, do their own extension, develop their own companies, find their own niches, craft their own organizations, whatever it might be, and they do this in a complete vacuum of institutional support or government funding, and I think it's all the more praiseworthy that they have gone as far as they have on this volunteer, individual, help-each-other sort of basis. It's now gotten to the point that they're so big that it's hard for them to continue to operate that way — somebody who's trying to run a dairy farm can't be on the phone mentoring other farmers all day long. And this is what it's come to now, because there's so much demand for farmers going into transition to go into organic, and the other concern, of course, is the Horizon dairy and people like that.

Q. Pretending to be organic.

CLARK. That's right, and if that's organic, then maybe we need to think of another word for what we're doing.

Q. On the other hand, when *Acres U.S.A.* started publication 30 years ago, there was not one professor in the whole system in Canada or the United States.

CLARK. Oh, I believe it.

Q. And now we are getting quite a few of them around that are walking the walk and talking the talk.

CLARK. It's very heartening and very encouraging to see that happening. I was at the SCOAR meetings this year — that's the Scientific Congress on Organic Agricultural Research — just prior to the big Eco-Farm conference there, and that was the first meeting of that group. It was over 100 people, and they were all brought there by the Organic Farming Research Foundation. And that was brought with, if memory serves correctly here, with actually a grant from the USDA, so there actually is some government recognition of organics, bringing together a bunch of people from across the country and even me from Canada — I am American, but I've been in Canada for a long time — and it was heartening, it was wonderful to see the breadth, experience, and again self-reliance, and the can-do attitude, and all that, from all of this range of academics and farmers and just a whole range of people.

Q. Now would you allow us to add the word "token" support from the government?

CLARK. I wouldn't use that word — but it probably is token.

Q. Maybe that's hitting a little below the belt.

CLARK. One of my concerns with managing government, and I'm not very good at this, so I'm not the one to do it, but I think at the moment government has gotten itself backed into a corner here — and this is government at the state level, provincial level, federal level — with a long history of active repudiation that organic could ever be anything besides tree huggers and granola crunchers, and these farmers — these "damn fool farmers" — they went ahead and proved 'em wrong, so now they have to either eat crow and admit that they were wrong, or continue to badmouth organics. I think the strategic way to deal with this is to find a way to let them get out of the corner.

Q. Let them save face.

CLARK. Yes, instead of hammering them deeper and deeper into that corner, which is my usual approach. So I think people with wiser heads than I, like Fred Kirschenmann or Chuck Francis or people like that, need to come to the fore and find a way to find common cause with conventional farmers, with conventional agribusiness, and with government, and let us get past this history, because otherwise we'll be Israel and Palestine, and we don't need that.

10

Teaching from Experience

A seeker with both academic and practical experience answered some of the most basic questions surrounding the pasture enigma.

David Zartman, Ph.D., is an innovator, a grazing specialist and an educator, in that order, if not in that order of importance. A native of New Mexico, he picked up his grazing knowhow before he even earned his several academic degrees, one of which was a Ph.D. he earned from Ohio State University. He still teaches contemporary issues — about animals and their use by humans — but management intensive grazing has become both a brainchild and a passion. The details are the substance of this conversation. The objective of his course work is to teach livestock production that is socially acceptable, ecologically correct, environmentally prudent and profitable as well. Zartman also supports the outreach and Extension programs in grazing and land use development. He is a member of the core advising team for freshmen in animal science studies at Ohio State University and the former chair of the Department of Animal Sciences.

Q. Dr. Zartman, as you teach your courses on grazing, are you calling into question some of the confinement-feeding practices that deny pasture to animals?

ZARTMAN. That's a yes-and-no answer. Yes, we do point out the contrasts between the two methods, but I don't denigrate or bad-mouth confinement systems. That is one set of management selections or choices out there. I simply focus on teaching a different choice.

Q. What is it that you teach, and what are the advantages?

ZARTMAN. I teach the concepts of rotational grazing and forage management. The first plus of this system is that you increase the productivity of your land by management intensive grazing. In most cases that increases profitability per acre — when compared to small-grain production, for instance. Another plus is that it is environmentally sound, and you develop a good, strong sod or turf in your land that reduces erosion and water runoff, with the benefit of improved water quality. It is a low-chemical-input system — people do use commercial fertilizers sparingly, but as appropriate, and probably no pesticides. This system encourages all the soil biota, including earthworms down to bacteria, which improve the health of the soil — its capacity to absorb organic matter and to recycle nutrients within the soil profile. With worm tunnels you increase water retention on the land, you increase the transportation of micronutrients from lower soil profiles up to the top in worm castings, and move organic matter into the soil.

Q. These are the things you focus on in your classes?

ZARTMAN. Another way of saying this is that I am trying to teach ecological farming. I use the grazing system as a prime example or method to improve the ecology of your farmland at the same time you improve profits.

Q. Rotational grazing is an important part of this, then.

ZARTMAN. Yes. Intensive rotational grazing is important.

Q. On what basis do you recommend this rotation?

ZARTMAN. It will be a function of the particular management system or species choice you are making. If you are looking at a non-lactating, non-growing animal — a mature ewe, mature cow, mature horse, whatever it is — then the rotation could be as seldom as every four days and still be consistent with the principles of which we speak. If you are looking at a lactating animal or a rapidly growing animal, where you want to encourage greater intake of a more constant character each day, then you want to rotate pasture every day. In the case of dairy cows, you probably rotate pasture every 12 hours, perhaps even making more frequent adjustments within that 12 hours.

Q. How important is grass to a dairy cow?

ZARTMAN. I'm going to answer that in kind of a teasing way. We speak of grass all of the time when we are engaged in this kind of discussion, but we really don't mean to exclude legumes. We really ought to say forage. We just get into the habit of saying "grass" kind of haphazardly, as if we are not thinking about legumes, brassicae and other high-utility forages. For the sake of this discussion, let's focus on forage. Forage is absolutely vital to the health of the animal from the following perspective: in contrast to harvested and preserved feeds, freshly grazed green forage is at its best condition, in most cases, for nutrient provision to the animal. One place you can really begin to put your finger on this is regarding vitamins. There are many vitamins that are quickly degraded in a harvested feed. Vitamin A is one nutrient in particular that is in very great abundance in freshly grazed green feed.

Q. Is it degraded in silage?

ZARTMAN. Silage, hay, anything that has gone through a drying and preservation process will have a deteriorated vitamin content.

Q. Have you made the connection between the availability of forage for animals and conditions such as Johne's disease that have become quite important in the United States? Have you made a connection between the availability of pasture and disease?

ZARTMAN. We find that cattle on rapid-rotational-grazing pasture appear to be advantaged health-wise in the following ways: the nutrient intake is of a much better quality, and the volume is usually adjusted by the animal. As long as it is good quality pasture, they will intake about 3 percent of their body weight on a dry-matter basis, which is what we really want for a lactating or growing animal. They do this in such a way that they become athletic; they are more vigorous and in better condition, although they may look leaner than a confinement animal. It is much like an athlete — it is not that they are deteriorated in health or condition because they are leaner, they are simply more athletic because of the exercise they get. Their feet and legs are much stronger and better. We find that animals in a pasture system last longer; their vitality and durability are greater than a non-grazing, confinement type of animal would experience.

Q. They get some exercise for one thing.

ZARTMAN. They get exercise. They get vitamin D from solar exposure, as well as the intake of the nutrients in the pasture.

They are vigorous, they are in the open air, and every night they have a clean bed.

Q. In *Soil, Grass and Cancer*, André Voisin pointed out that one of the deficits in non-pastured maintenance of cattle is copper availability, and this is being related by some research at Cambridge to the appearance of Mad Cow disease. The question is, how available are the nutrients in a confinement-feeding system, where you have to get all of the different nutrients into a mix of some sort, as opposed to what nature provides out there in the pasture?

ZARTMAN. You get yourself in trouble if you think that pasture provides all of the micronutrients in appropriate amounts. You cannot expect that. How do you solve that dilemma? First you have your soil analyzed so that you can properly feed the plants. You need to have your forage analyzed so that you can properly feed the animals. We find, for example, that just about all of southern Ohio is seriously copper deficient in plant analysis. So your mineral supplement, and I'm calling it a given — you will have a mineral supplement whether it is a grazing system or not to meet your micronutrient requirements — and copper would be one that would require special attention for some areas.

Q. In other words, this is dependent on your geographical area and the quality of the situation where you find yourself?

ZARTMAN. Yes. Species is also important in your ascertainment of nutrient needs. Sheep would probably not need supplemental copper here in southern Ohio, because apparently they don't have as high a requirement as cattle. Maybe they are better at extracting it from their forage. There seem to be differences even among cattle. Work at the University of Kentucky indicates that Jerseys don't need the same copper supplementation as Holsteins do on the basis of body weight.

Q. The perception that too many people have growing cattle is that if I have a pasture, it's out there, and I don't have to do anything with it.

ZARTMAN. That's quite wrong. That's why we call it management intensive. What you are doing is taking yourself off the tractor seat and going out on your feet and turning your head on instead of your tractor engine.

Q. What kind of fertility approaches do you teach?

ZARTMAN. We expect to feed the plant. My guiding statement to students when we begin is that we are teaching them to be grass farmers, and by grass I mean forage. The only thing you have to figure out beyond producing a heck of a lot of good quality forage is what kind of harvesting machine you want, and I recommend you use an animal.

Q. That animal also distributes fertility, does it not?

ZARTMAN. Exactly. In fact, when you build a system like this it tends to be pretty much nutrient/budget balanced. What you buy in the form of supplements to bring onto your farm is pretty well balanced with whatever leaves your farm in the form of meat, milk or eggs. With about an equal amount coming on and leaving, your farm is in a balanced condition. Within a few years of a management intensive grazing system being installed, you will come to what is a naturally balanced system of animals consuming the nutrients and recycling them. As the animals harvest, they fertilize, they recycle every 14 to 16 days in the early part of the year and the fall, and every 30 days in the summer. They harvest and fertilize.

Q. In a rain-belt area the rain will do the flushing of the ground. You won't have to irrigate it.

ZARTMAN. But even that is an open-ended question here in the upper Midwest. We have this dry spell in July and August when everybody's lawn goes dormant, and so do the pastures. The lawn managers say that an inch of water a week in those times will keep a lawn from going dormant. I think that would also work for a pasture. That's one of the next chapters I want to move into in my work, application of strategic irrigation of limited acres to serve as a reserve to get through the summer period, while the rest of the pastures rest.

Q. Is that an additional course to the one you are teaching now?

ZARTMAN. Someday it will be.

Q. Fleshing out the course that you are teaching, what would you add?

ZARTMAN. In the beginning, don't be overwhelmed by the complexity — just begin in a simple manner. Just begin. After I have talked to my students about becoming forage producers, I tell them they are going to learn, through our practical work in the laboratories here, that they can go home and do this work at the end of the quarter in a simple, applied way. As time goes on, they will become more sophisticated and more knowledgeable. My point is, don't get blown away by the complexities, because the fact is that you can do this very simply in the beginning, and it will work.

Q. How do you gear the carrying capacity to the pasture?

ZARTMAN. There are a few rules. The first rule is that if the pasture is of reasonable density and composition, you can expect 400 pounds of dry matter per edible inch on a per-acre basis.

Q. Over what period of time?

ZARTMAN. For that we have to look at our rotational duration. This time of the year we are rotating on a 14-day basis. Every 14 days we are back where we began. We have a 14-day supply of feed. Based on how many inches of forage there are, say 8 to 10 inches, and the animals are going to eat 5 or 6 inches out of 8 inches. If it is 5 out of 8 inches of grass — and that's a pretty reasonable estimate — the animals will consume 5 inches of 400 pounds of dry matter per inch on an acre basis in 14 days. Then you look at the amount of acreage you have and divide by 14 to know how much forage you have each day. Once you know how many acres you have plus yield per acre, that will tell you how many pounds of forage you've got on a given day. You divide that by .03 — the 3 percent figure — and that tells you how many pounds of animal you can feed. Then you decide what kind of animal you want.

Q. If you have a dairy cow, it will be significantly different from a goat herd. Do you have those equations figured out in your course?

ZARTMAN. Yes. It is just pounds of animals. Your growing or lactating animal would be 3 percent, and a mature non-lactating, non-growing animal would be 2 percent. It is just a matter of animal body mass that you are feeding. These make very good rules of thumb, and they work very well.

Q. What other rules of thumb have you uncovered as you have developed your course work?

ZARTMAN. Be flexible. Don't marry yourself to your initial plan, particularly in the beginning. New graziers will miss two points. Number one, they will not expect as much feed as they are actually going to have. They will have set up their rotation expecting that they were going to harvest less feed than actually is there. That creates problem number two: there are not enough animals.

That can lead to a third problem: if they stick to their original plan, their pastures will get overly mature before they get to them toward the end of the rotation, and then they are caught in a dilemma. They say, "I've got my system, and I have to follow it," and then all of the pastures get out of sync. Otherwise they have to say, "I have to change my plan and go back where I started and make hay out of the pastures that got ahead of me." My advice is to be flexible and continuously monitor what is going on. As you go out each day (or some period of less than four days) to move your animals, look at the animals and see how they are doing. Make an appraisal of the animals' body condition, grazing habits, and their behavior — they will tell you whether things are right or wrong in large part by their behavior. You have to be ready to change your system accordingly. There are other factors, such as climate and weather. We've had 3 inches of rain in the last few days, which means we have grass growing like crazy now, and we have to be flexible and deal with it. We also had a little dry spell that put some of our tall fescue in the condition of wanting to go to seed, which means you have to get out there and do some clipping. You don't want a lot of seed heads out in the pasture.

Q. What do you do to prevent overgrazing in deficient periods?

ZARTMAN. If it hasn't been raining, and there is no humidity, your grass quits growing; if it is hot, it may go into dormancy. You are going to have to change your stocking rate right then to prevent overgrazing. That means you move more frequently. You stock the animals at the same rate, but for a shorter period of time. Basically, you decide how much feed there is and divide by 3 percent, and that is how many pounds of animal you can feed right then. You have to have other pasture area to go to.

Q. If you don't have the other pasture area, then you have to find some other feed source.

ZARTMAN. That's right, you supplement, or you change the number of animals that you have. If you are a stocker operation, go and market some animals and turn them into cash rather than buy feed. Those are the kind of adjustments you have to make.

Q. Economically, it works out to the best use of the land with the best profit if you go in for a grazing operation — is that what you are saying?

ZARTMAN. Exactly. We have a pretty nice set of data out of the University of Illinois for the last two years. They have been doing field studies of cooperating farms, and the data are pretty consistent with Ohio, as an example. They report that for small grains their net profit per acre is unlikely to exceed $80 and will most probably be in the $70 range. With today's grain prices, if it weren't for the deficiency payments of the government program, it would be a negative problem. With a management intensive grazing system for dairy, you can readily demonstrate better than $600 per acre net profit. With fat cattle, or non-dairy cattle, you should readily realize more than $300 per acre. With sheep, it would be somewhere between those two numbers, but it can get pretty close to dairy. With horses, it is difficult to make any estimates, because it all depends on what you are selling and what your market is. We do get dairy farmers who are reporting numbers much higher than $600 per acre; some are over $1,000 per acre.

Q. This is just dairying, no fancy business, no BST or anything like that?

ZARTMAN. I wouldn't want to exclude BST, because I don't know who may or may not be using it. Most of the graziers tend not to use it, but I can't really sort that out. Most of them happen to be seasonal dairies as opposed to year-round. That's because once they have learned to play with nature in the growing period, then step one is grazing, and step two is to be seasonal, because that is also consistent with a particular marketing philosophy. If your philosophy is to maximize your milk income, you would want to be seasonal, with fall freshening. You will spend four months in confinement before you get your pastures working and you chase the milk price. On the other hand, if your philosophy is to reduce the cost of production, then that dictates you are going to be basically a total grazier and maximize feed on the lowest-cost feed you can get, and that is pasture.

Q. Your basic training and Ph.D. was not in rotational grazing, was it?

ZARTMAN. The short answer is no, but the long answer is yes. I was born and reared on a progressive dairy farm in New Mexico with irrigated farming. We were very keenly focused on pastures and pasture management. My father was a cooperator with the New Mexico A&M (now New Mexico State University)

Experiment Station. We were a cooperator farm, regularly taking data and doing a variety of trials for pasture. We used all variety of pasture management and design that was thought to be interesting at the time, and we held regular field days every year where people could come in and see what was going on. We used rotational grazing, but not as intensely as we know it today — we did not have the fencing technology in those days. Toward the end of my youth, we did use electrical fencing, but we were not pushing it as intensively as we are now. We just didn't have all the nifty equipment that we have these days for rapid-rotation grazing, such as poly wire, step-in posts and all these conveniences we have now. So I had training in grazing from the time I was a baby boy, but my formal training in academia was on the side of genetics.

Q. Apparently you have picked up on that youthful interest and reinserted it into the educational system.

ZARTMAN. I spent a year in New Zealand shortly after I became a professor, and that was a life-changing experience. I did a post-doctoral there with the National Institutes of Health. I observed how intelligent and practical those farmers were. I came back to the United States very much reoriented in my approach to being a professor and a scientist. Even with that, I didn't really return to grazing until I was here in Ohio, and in 1987 the East District county agents called me over to their annual planning meeting. I was chair of the Dairy Science Department at the time. They said they needed my help with an idea they had. The whole economy of Appalachian Ohio had taken continuous hits through loss of jobs, loss of industry, consolidations of schools, and tax bases declining. They thought nothing was working and believed that dairying might be able to work in the region, but they needed help to figure out how to do it. All at once everything sort of clicked — New Zealand, similarities, my background experience. We set up a five-year project at one of our branch stations.

Q. Was this oriented to smaller or family-sized farms?

ZARTMAN. You can make that interpretation, but it is not a limitation.

Q. It is not a limitation, but you are not talking about a 5,000-cow dairy operation, either.

ZARTMAN. Probably not 5,000, but 1,000 might work. There is a Wisconsin dairyman who is doing 1,000 head. There are a number of farms in New Zealand that are in excess of 500 cows that do this. There is another dairy in Wisconsin that is over 400 head of grazing cows right now.

Q. What would you put as the top figure for a grazing operation?

ZARTMAN. It would probably be about 1,000. The question is: how many milking parlors do you want to build? Some places walk the cows more than a mile to milk them, but that is quite a trek, and I would prefer not to go more than a half mile. Then you would ask how many cows can you graze in a circle that has a radius of half a mile.

Q. You relied a great deal on your New Zealand experience to provide the framework for your course work, I gather.

ZARTMAN. Exactly. A lot of people say that Ireland is also a mecca for grazing. I haven't been to Ireland, but New Zealand generally is the place I point to when someone asks how to reduce the cost of animal production.

Q. How successful is this proving to be in Ohio?

ZARTMAN. Quite successful. The thing that amazes me is how low the adoption level is. That is a little bit of a university problem, in a way. We had some really sound paradigms from the '50s and '60s that we taught people. We learned these paradigms from our research program, put them into our Extension format, and taught people these paradigms so well that they can't let go of them now. They are all in the context of dairying, and there are four basic paradigms. The first is that the more milk you can make a cow give, the better. The second paradigm is that anything you can buy to get another increment of milk out of a cow, buy it. The third paradigm is that each cow should be managed as an individual, so you can optimize input relationship to output. The fourth paradigm is that manure is waste — get rid of it any way you can. Our input costs on dairy have gone up four times, while the value of milk sold has not even doubled.

Q. You hear a lot of criticisms of the industrial model being fastened onto a biological procedure in agriculture. In other words, they want to treat the dairy like a factory.

ZARTMAN. I don't think that is necessarily a wrong way of thinking. The question is, what kind of factory is it? It's a biological process, not a mechanical process.

Q. That's the point, isn't it? How well are your teachings being received by the student body?

ZARTMAN. I can give you an indication that is probably better than me just saying, "It's great." The first time this course was taught was five years ago. We had done our five-year project and produced a bulletin from the Ohio Agriculture Research and Development Center that reported our work, and it was igniting a lot of interest around the country. One of our county agents who had become a leader/specialist in this area of management intensive grazing said he would voluntarily come in to teach if we got a class together for it. We put out posters and word of mouth in our animal sciences department, which had just gone through a restructuring so that all the animal departments were merged together into one unit, and I was chairing the department. We got 10 students together the first time it was offered, and we taught it just as an experimental course. A year later, he tried it again and only got three students, so we canceled it that time. Then our volunteer retired and went to Mongolia. In the meantime, I had made my decision to step down as chairman. I had this fire in my gut for grazing and ecological farming, so I told my students I was going to pick this course up. Six students came forward, and we did it as an individual group studies class. Those six students did wonderfully. In fact, one of them wrote an essay that was submitted to the Forage and Grassland Council essay competition, and he was the national winner.

Q. From then on it flowered out?

ZARTMAN. Last year it went from six students as an informal course to being a formal course installed in the curriculum and listed in the catalog, and there were 38 students enrolled. This year we have 41 students.

Q. You can't develop a course at a university by waving a magic wand. It took some putting together to get a course structured for an entire semester, didn't it?

ZARTMAN. It took a lot of work. We are also building a network of folks around who are supporting this even more vigorously. I want to emphasize that some of the difficulty in getting this going was the sheer resistance of those old-paradigm teachers who think that a confinement system is it, that big and confined is the way to go. I guess you might call that factory farming. There was a lot of resistance to grazing from my colleagues who felt it was just so old fashioned and that it wouldn't work, particularly with western Ohio's high-value grain ground. They felt the dairy industry just wouldn't go in that direction. The beef industry was more amenable to it, because they were already grazing oriented, but not intensive. So many of them are small operators, and they live on off-farm income, and the cattle are more of a hobby they are willing to subsidize with off-farm income.

Q. An awful lot of your cow/calf operations are sort of hobby operations.

ZARTMAN. There was a lot of resistance to this when I started the group study course, but it came out so well for those six students, and we got profound support out of people like Bob Evans, the restaurateur, farmer, rancher, entrepreneur and enormous supporter of management intensive and year-round grazing. He is so strongly in support of grazing, and he puts his money where his mouth is. With that kind of support, people stopped fighting me on the idea. They thought I was out on a limb, a fringe operation that would never be mainstream, but all at once they realized the folly of their thinking and began to support me. A number of the faculty started to advise their students to take the course and learn about management intensive grazing. Now the students who have taken the course are recruiting among their peers. This is an elective course, not a required one. When you get this kind of enrollment in an elective course it is very gratifying.

Q. Do you see these kinds of courses making headway in other universities that you are familiar with?

ZARTMAN. Mostly they are just beginning. There are university departments of agronomy or horticulture that teach forages and maybe even grass production and management, but they are not teaching grazing from the animal perspective. I think it is up to the animal science departments to do that, and there are not very many doing it yet.

Q. Maybe your experience will start inoculating the idea in other places.

ZARTMAN. That's my dream. I guess you are going to leave the door open for me a little bit here. I have another course where I tell students that I believe that the days of what is generally known as concentrated animal feeding operations are numbered in the United States. I'm building my case on three observations. One is the intensity of the environmental movement calling for improved standards in air and water quality. You are hearing now about TMDLs — total maximum daily loads — as far as state-wide water quality standards. The EPA wants every drop of water that is publicly located in the United States to be fishable and swimmable. We are way short of that, and they claim that the animal industry is a basic violator of these standards.

Q. So that massive feedlot is a sunset operation from that point of view.

ZARTMAN. Yes. The second observation concerns the animal rights movement, which is making very sure and critical inroads into the concepts and the methods of animal production and the humane treatment of animals. They used to try to do it pretty much head-on through legislative methodology, but that is pretty slow, hard work. Now they are going to the consumers through the food industries and orchestrating change from the food industry back to the farm through consumer pressure. My third observation is the sheer deficiency of labor willing to work on these operations at an affordable price, that is, labor problems are an ever-increasing difficulty for concentrated animal-feeding operations.

Now, if you were in that business, and you had to deal with those three pressure points, and you could look and see that in other countries you wouldn't have to deal with them, wouldn't you move? They are already moving to South America and Mexico; some are moving to China and former Soviet bloc countries such as Poland and Yugoslavia, somewhere where there is enough political stability to gamble on a concentrated operation.

Q. Those are the reasons you think these companies are moving?

ZARTMAN. I think it will take only one human generation, and we are already well into it. A human generation is 45 years, so I am saying that in about 30 to 35 years many of our concentrated animal feeding operations will have relocated outside of the United States.

Q. They will pollute Brazil or Mexico or wherever it is until they stop them.

ZARTMAN. Why do I say this? I am saying this on the basis of what is going on in our public school system. Many of our teachers are young, because there is so much turnover in school teaching. These young teachers are green oriented, they are environment, ecology, and food-safety oriented, they are anti-hormone, and they are teaching the children consistent with what they themselves feel and believe.

Q. This is arguing the organic case then, isn't it?

ZARTMAN. Absolutely. That opens the door. If I am going to be condemnatory of these concentrated animal-feeding operations and say that they are going to be moving out of the country, then what do you do to fill the void? You graze. You go into forage-based systems. Who will support that? It is environmentally appropriate, we already talked about what this system does for the environment. Number two, it is appealing to the public — you might call it socially acceptable. Animals on pasture usually look nice if they are healthy and happy.

Q. We even have people who have what I call decorator cows, maybe three Longhorns on their corner lot.

ZARTMAN. Think back not very many years ago, before they got into their special market situations up in the New England states. Vermont used to subsidize milk at a dollar a hundredweight on the basis of tourism. They did it to maintain tourism; they had to keep those dairies going up there. Now they have a compact, so they get it through the grocery stores, but it is the same thing. They keep their dairy farmers in business. The third thing is that it is profitable. I say that in the context that there will be niche marketing. There will be a certain portion of our population that is willing to pay the premium for locally grown products, in which they have more assurance or confidence in the quality and nature of that food. I don't think that will be the bulk of consumption in America, but it might be called "organic," or it might be called "naturally grown." There are different phrases in use today — it might be part of the free-farm program that you may have heard about, which is being promoted by the animal humane associations. Maybe up to 30 percent of our consumer dollars will be spent on food of that kind.

Q. That's a big jump from what it is now.

ZARTMAN. I was at a conference in Pennsylvania called PASA (Pennsylvania Association for Sustainable Agriculture). I was speaking at their annual meeting, and there were about 1,200 people there. The groundswell of thinking at that meeting was organic, and some nice data were reported. They said that consumers will comfortably pay up to about 10 percent more for food that is organic before they balk.

Q. Especially as they become alarmed by things like bovine spongiform encephalopathy, or Mad Cow disease, and things such as that. There are a lot of people who are shifting entirely to making local purchases and private treaties with the farmers themselves.

ZARTMAN. Then you can also throw into the mix issues such as neutraceuticals, foods that have naturally increased disease-fighting or health-promoting features. Conjugated linoleic acid (CLA) is a good example, because it is about four to five times higher in a grazing animal compared to a non-grazing animal. It is probably the most impactive natural cancer-fighting food known right now. CLA is found as a result of ruminant animals eating green forages, so it is quite high in the milk of a grazing cow or the meat of a grazing sheep or beef.

Q. This information is slowly seeping into the consciousness of the consumer.

ZARTMAN. There are some other things, as well. There are specific fatty acid profiles that are healthy in grazing animals and different in composition from non-grazing animals.

Q. Other than yourself, how many people are teaching this information in the university system?

ZARTMAN. I don't really know. I guess I would say that it is still only a few. The news about CLA — that's pretty broadly based in the agricultural literature, but is it getting into the dialogue of the classroom? Probably not much, and that's because not very many classrooms talk about grazing. They are mainly talking about nutrition *per se*, just nutrition, and most of the time that's in the context of a balanced ration in a feedlot system or a confinement system where you have a perfectly balanced ration. By the way, that is the key to a lot of resistance to grazing, because from a nutritional point of view, in their ideal world, they would provide a balanced diet every day. With a grazing system, you don't know what the intake is, so you can never arrive at the peace of knowing that you've got a balanced diet. You don't know what they ate or how much they ate.

Q. But it is a free-choice system, and the animals are very likely to eat what they need if it is available.

ZARTMAN. That's the assumption that you make, but the nutritionist doesn't like those assumptions; he wants assurances.

Book Two

The Graziers

11

Putting It All Together

The seekers and the grazers have so much in common, separation by categories becomes almost impossible. The true grazer can be isolated in case reports, each case representing a biography and a model for ranch management.

Tom McGrady has a cow-calf operation in southeast Texas, some 45 miles north-northwest of Houston — "on the edge of the piney woods," as Tom McGrady puts it, "in a 45-inch per annum rainfall area." Cartographers and geographers would define the area as sub-tropical.

The ranch has a carrying capacity of about 250 head of cattle, the venue being grass — forage, to be a bit more technical. The original animals are Limousin, from an Angus base. McGrady hasn't set out to make the herd purebred, but he has been line breeding since about 1990. "We do have some purebloods," McGrady said. "We have some three-quarters that are Angus-based, bred up using a black pureblood bull rather than French bulls." These bulls already carried the Angus traits. McGrady's herd bulls are out of the best and most productive cows. The sires of those bulls are the ones that combine the character traits necessary for producing animals that will meet his goals in yield and quality. "We are developing cattle that consistently produce choice, grade 1 carcasses." To date, "We've been moderately successful," McGrady said.

Given enough time, according to the precepts of line breeding, the art and science can create a new breed (and has done so), preserving the best of the best in genetics. "Line breeding preserves and builds," according to McGrady, assuming a timeline in harmony with the life span and heritage of a family that often attends the development of herd and breed building.

A project now underway was hatched when Tom McGrady and Indiahoma, Oklahoma cowman Jim Lents exchanged knowledge of genetics during talks that had no defined beginning and no likely end. He wants to create a Hereford-Senepol gene pool, a heat-tolerant maternal composite of a smaller-type cow with ample flesh. The objective of the Hereford-Senepol mix is to combine the heat and humidity tolerance of the Senepol with the type, body composition and grazing capacity of the Hereford. While on a trip to purchase a few Senepol cows at Teddy Gentry's Bent Tree Farms sale one April, McGrady saw the Trask Polled Hereford cattle, a gene pool of cattle line bred for over 70 years to perform on grass alone in the heat and humidity of the Southeast. The sight of Polled Herefords as they "ought to be" prompted him to purchase a nucleus of cows to be expanded through embryo transplant. A Hereford bull of Anxiety 4th breeding, a product of 22 line-bred generations, was selected to mate with these select Hereford and Senepol cows. Through line breeding, the results will be homozygous.

With the line breeding approach securely in place, McGrady opts for grass, fenced paddocks and a highly refined rotational grazing system. "Essentially," explained McGrady, "other than breeding season, we combined the cows into one herd for grazing." He has about 45 paddocks, each a product of experimentation, development and care. He has rejected the monoculture approach and its debilitating drain of fertilization. He has relied on managing grasses in a green-leaf stage. Nothing, absolutely nothing, can replace good turf, the greater the variety, the better, McGrady said.

The area in which he ranches is unique. Soils can be defined as everything from sugar sand to black loam. It takes many species of grass to hold these soils together.

Many cow-calf operations are owned by off-farm income earners, often by dilettantes who have still to learn quite a bit about the bottom line. They like the brightly painted toys that equipment factories have to offer. They seem to follow the model of taking a high-priced calf, feeding it with a high-priced ration, calling the vet for high-priced vaccines, then selling at a loss.

Tom McGrady has almost a Ben Franklin attitude about a penny saved being a penny earned. He hasn't wormed any cows in almost a decade. He relies on the herd's "body terrain" to ward off viral, bacterial and insect health destroyers, that terrain being managed with genetics and nutrition.

"Essentially," explained McGrady, "we got down to one herd of cows we rotate, keeping the very best bulls for the very best cows. The sires of those bulls are the ones that combine the best character traits." Genetically this means staying within the 50 percent range decreed by the line-breeding model.

All the above is merely a prelude to the real message from McGrady and east Texas. The old saying "all hat and no cattle" has to be modified, because a sprawling herd may not be the answer to a bottom line.

Some few years ago, McGrady had some 250 animals on grass. Adjustments required by the economic conditions surrounding a divorce caused him to reduce the herd size to 100 head, its present strength. McGrady raises his own replacement heifers and bulls. His logic towers above most of the coffee klatch billingsgate at the local restaurants.

When ranchers do not raise their own replacement heifers, the whole year is focused around weaning and selling a calf. Grass management calls for a different sequence. Cow condition dictates weaning dates and grass conditions dictate stocking rate. A weaned calf can continue to be grazed, then sold when needed to cut the stocking rate so that the remaining herd has plenty of grass. You can put tallow on the cows that way when they're dry. This cancels out the expense for supplements.

Many ranchers simply do not know how to grow out animals on grass. They stuff a lot of feed in them. They believe it is cheaper to buy a replacement. As a consequence, the quality and genetic integrity of the incoming animal is often compromised, rendering them unable to sustain the quality, type and body composition needed in grass based operations.

"I like to wean my calves on a hot fence. This means I put the herd in a grazing cell center, cut the cows out and put them back into the paddock and put the calves in the paddock in front of them — this lets the calves selectively graze fresh grass. I keep moving them, with the calves always grazing in front of the cows," McGrady explained.

When he needs to cut his stocking rate, he has short or long yearlings, calves already weaned on grass that are available for sale and stocking reduction, all with minimum stress. McGrady tells of calves sold off without even one sickness for the next 90 days.

The potential for finishing calves on grass is great, according to this Texas rancher, the key being matching genetics and grazing management. This means cattle have to be genetically capable of finishing out on grass. The selection process makes the determination. The Hereford, a longtime resident of this country, built its dominant position during the first part of last century on grass. The Hereford-Senepol cross McGrady is developing is heat tolerant, true beef animal that can finish on grass. As numerous studies have shown, pour a lot of corn into an animal, and it gets too fat — with the wrong kind of fat. Cows on grass have the right kind of fat — "and I believe there is a big market for such quality meat," McGrady said.

Unfortunately, the institutional arrangements for doing business do not accommodate the producer of quality, for which reason the consumer circuit has to become the wave of the future, at least for the producer of meat above commodity standards.

Merely seeing to the genetics that can thrive on grass is only half of the equation. There has to be grass management geared to keeping forage in front of the animals at all times.

The nemesis of the Texas cowman — indeed, the cowman west of the Topeka Meridian — is drought. "We like to wean our calves early in the case of drought," McGrady said. "We constantly have a drought plan. We have critical dates. Without enough moisture by these dates, we know to implement the program."

The key to cows on grass is to always have enough forage to meet the feeding requirement. Extra grass is there to grow out calves. Yearlings can always be moved to answer the situation. The stocking rate can be cut as needed in half with timely warning, the product of a proper grazing plan.

There are authors that intone the theory of herd numbers. Some ranchers keep hammering the soil even when rains come. "Sheep-footed" soil won't insoak rain, a defect answered via timely rotation. In any case, a rotation can be lengthened to buy time. When rain finally arrives, it will touch off the growth of grass.

There is an old saw in agriculture to the effect that "this ground isn't good enough for a crop, so I'll turn it into pasture." This, of course, is errant nonsense as far as the grazier is concerned. Growing enough grass is only part of the equation. That grass has to feed microbes as well as the cattle. Grass has to construct a shelter for the microbes in the soil. If the underground construction job isn't accomplished, there won't be enough grass to feed the animals. If husbandry is extended to those billions, even trillions of unpaid workers, a turf of grass will eliminate bare soil. Graze such grass — not into the ground, but reasonably — allowing food for the animal, sparing a little to be tramped down into the soil for microbes to break down, with adequate rest between grazings — that to Tom McGrady is the bio-equation for grass-fed and finished cattle. Urine and manure are both a byproduct and a fertilizer, one that can't be replaced economically with factory salts. None of the above can be accomplished with six or eight paddocks, according to McGrady. It takes upward of 25 or more paddocks.

Thus, genetics and phenotype depend on the land. No college scientist has ever been able to genetically modify a crop or an animal to withstand starvation.

Starving an animal to the bone is an absurdity. Starving a pasture by grazing it to the soil itself dries the upper inches by offering them to sun and air for baking. When this happens, all that microbial activity dies. Rain will not restore life as easily as devoutly wished. Life will build back, but the trail back from needless debilitation is never easy. Leave turf in place, and although the drying process will still take place, the road back is almost magical.

Rain-making has been around since the 1940s and even earlier, but as a practical matter, few ranchers believe they can order up rain. They get what nature decrees, no more. But they have a great deal of control over how much moisture they keep. It can be stored in the soil, but it takes organic matter to accomplish that purpose.

It has been proposed, of course, that an irrigated paddock be used to crutch over a dry spell. McGrady does not see that proposition as economically viable, at least not for his Texas acres. Whether the idea is of merit in other areas of the country must be evaluated "on scene," not in the abstract.

A cut in the stocking rate and increasing the rest period both trump irrigated acres for the grass crop.

"One year we had a drought, with less than 40 percent of normal rainfall," McGrady recalled. "We only made three cycles for the whole year — an average rest period of 120 days per paddock. In our country, that's a long time! We cut our numbers, we weaned our calves early, but the cow numbers were not cut drastically — 10 to 12 percent."

Cow herds usually have a top one-third, a middle one-third and a bottom one-third. Usually the bottom third isn't making any money. Under conditions of drought, that bottom third deadweight is counter indicated. They have to be moved out.

The idea of a cow herd being a fixed number — the effort to avoid the "all hat, no cows" sneer — opts against flexibility, and yet the herd number has to be feasible. Raising your own replacements is a good way to add or subtract, and to maintain the genetics, as well.

McGrady believes in raising his own bulls, not only to avoid being caught up in ad hype, but also because it is a given that there are people who misplace their integrity. Many cowmen become Expected Progeny Difference-result oriented and try to make use of genetics from an artificial environment in a range setting. Such genetics are not suitable for herd improvement, McGrady opined. This demurrer is less philosophical than economical.

"Look, if I could buy bulls that work in the real world, on grass, for the purpose of raising replacement heifers, that would save me a great deal of time," McGrady summarized. The people who raise the bulls being advertised have only marginal success. "They do not get the spotlight or blue ribbons. You really have to scout the terrain to find them."

Texas cowmen say that there are a lot of overgrown 4-H project ranchers "out there." The paradigm that sees a worthless ribbon as an objective rather than a bottom line does not comply with practical reality. Egos tend to collide with practical ranching, leaving the real world out there as an apparition to be shunned.

Why is this? McGrady has his theory: "In other parts of the world where there are no auction barns, weaning weight is irrelevant. They'll wean a calf at four to six months of age because they can run that cow dry more than half the year on little or nothing. She can make it on quantity of feed, drought stress notwithstanding. The quality grass is reserved for the calf. Here, we don't do that. We put out fertilizer to increase carrying capacity."

McGrady recalls fertilizing the entire place, running about 150 cows before he had completed the Stan Parsons Ranching for Profit School, Albuquerque, New Mexico. Parsons was associated with Allan Savory for a time, and much of the available insight was and remains based on the "brittle environment" — a growing season of about six or seven weeks a year, with grass rationing the chief occupation most of the year. The limiting factor is rainfall. With rainfall in a brittle environment, grass will grow and stay.

In a high-rainfall environment, grass can elude management, getting seedy and stemmy and wet. Most cattle operations in the United States are in high-rainfall areas.

The one-third, one-third, one-third analysis made a real fact of life jump off the spreadsheet for McGrady. "I was spending about $24,000 a year on fertilizer to raise all my grass and my hay for 150 cows. I found I could cut out $20,000 of that and simply fertilize for hay. But I cut my stocking rate to 100 cows."

In short, the 50 cows above were costing $20,000 a year. The county agent argued that McGrady ought to divide by 150 rather than invoke the three-part division. But a little pencil pushing revealed that he could have run 50 cows and not even fed any hay.

Later, with paddocks in place, he could increase the carrying capacity to 180 head with no fertilizer expense. Then, to manage grass under conditions of higher rainfall, when the grass continued to elude management, the extra grass was baled for hay.

This rainfall hay is different from fertilized irrigated hay. Here the only cost is the baling, all of it harvested from extra paddocks.

By moving in the right direction, McGrady found himself running more cows than he did under the trail drive system made famous by movies. The soil and the symbiotic relationship with the cow turned out to be a better answer. Soil, in turn, means microbial activity, organic matter — in short, a carbon and nitrogen cycle working naturally.

Grass, indeed, is the forgiveness of nature, to quote old-time Kansas Senator John J. Ingalls. When grass is in a green-leaf stage, cattle will eat it. It will grow back into an equally green leaf. When that green leaf is cut and rolled up into a bale of hay — $7 to $8 being the cost — a winter's feed comes to about $25 worth of hay, not the usual $100 for feed. That makes extra grass an added-value product. Thus, the choice becomes clear: Put up more hay or run more cows.

In eastern Texas, there are years when cows do not require even one bale of hay.

Hay storage for an emergency is not necessary, according to McGrady, not for his area. He has paddocks well ahead of the rotation for emergencies. One extra day of grazing equals 45 days of extra rest for grass in process in each paddock.

The equinox is identified as June 20 in the Northern Hemisphere. After that, the days of sunlight start getting shorter. If rain arrives after July 4, it will not deliver as much grass as rain before June 20, or even during the time line between June 20 and July 4. This projected absence of quality grass tells the rancher that he is required to adjust stocking rate, quality and quantity being foreordained. If the season is extremely wet, the cattle can trample a lot of forage unless temporary fencing is used to adjust paddock numbers. A move each day to new grass is indicated.

Carrying capacity is more static. Stocking rate must be adjusted accordingly. You have the axiom — a cow to every two acres each year, or some such equation — but axioms do not even out seasons, rainfall, temperature, drought or excess moisture. Averages be damned — if a cow goes 60 days without grass or hay or food, she'll die.

The winter season is the lowest in carrying capacity. Feed available has to be matched to that carrying capacity. This means dry, fat cows with tallow, so they can draw on that fat for winter survival on hay.

A mature look at simple facts and figures reveals that the people who have profit lines in agriculture are usually the feed salesman, the fertilizer people, the biologics and veterinary practitioners, the import dealers. Real ranching requires the use of grass as a value-added product, and a general disdain for the expensive toys that sponsor four-color ads in slick magazines.

"Minimal inputs" define a goal equal to others in husbandry. Even grass-mowing equipment is optional. The correct water system isn't. Cows actually do most of the mowing in the paddock.

In the past, this cowman has calved in February and March, a reasonable target for east Texas. "I'm looking to move it to December and January," McGrady said. "I like to wean my calves by Labor Day. If I can do it by then, I can rely on plenty of good grass into September and October. A dry cow will get fat on that good grass. If she gets fat, she'll hold her own in November and December. If she calves at that time and still has fat, she won't go down very much before the green grass arrives in spring."

Cool-season grazing is available in east Texas by the middle of February in most years.

Nature's own diversity is always operative. Cool-season grasses seem to move in and assert themselves to fill a niche, grasses that were never there before paddock management. At the Houston latitude, winters stay relatively warm most years.

Direct marketing is devoutly to be wished, according to many cattlemen. The same question posed to cowman Tom McGrady served up an answer that needs a bit of assaying by all those grazing cattle on grass. "It's something Jim Lents and I have talked about. You know where we should be sending our Beef Check-off money? We should be spending it to put small packers into business. They need efficient equipment so they can compete with the big packers. We have to process our cattle."

Allergy victims are demanding grass-fed beef, and physicians are seeking and sometimes finding supply sources. Unfortunately, most cattlemen have only looked at the idea, few have tried to put together fellow ranchers for the purpose of getting paid for as much wealth as they produce. The prospect of selling a 1,200 pound grass-fed steer — even a 900 to 1,000 pound grass-fed steer — at, say, $2 a pound for the carcass weight, that puts $1,000 on the collection side of the equals sign. That equation with little or no inputs would allow the grower to run half as many cows — a pretty solid living! Debt retirement beckons under such a circumstance. Paying for feed, fertilizers and equipment is a poor option.

Grass, in the opinion of Tom McGrady, does a fine job of balancing minerals into good pasture-management equation. He won't criticize the mineral box, but he feels balancing micronutrients is too exacting, often causing more trouble than it solves.

"Give a cow plenty of grass and let her pick what she wants. The few cows that won't respond seem to be in the bottom one-third of the herd." Ruthless culling is indicated. Moreover, to make genetic improvements it becomes necessary to identify longevity and productivity. That means weaning 10 or 12 calves out of a cow in a lifetime. Cows that do not require inputs will do that.

The highest cost of a cow herd is depreciation. Take the depreciation out of a herd, and the books keep themselves. "You don't lose a quarter, there is finishing capacity, fertility, good udders, fleshing ability, depth of body to accommodate roughage, raw material for tallow, meat and muscle. All in a low-stress, low-maintenance environment," McGrady said. A vet told him, "I can give all sorts of advice, but there is no substitute for plenty of good grass."

McGrady has experimented with not grazing one-seventh of his paddocks for one year, this to permit microbial workers time for some creative evolution. This rapport with the biblical approach is no accident. The lure of rested paddocks for winter feeding can be imagined. The feed can't be called high quality, but high volume it was and remained. Cows ate it, McGrady said, and trampled some of it into the soil for microbial consumption.

High-successional grasses — blue stem, eastern gama grass, Indian grass, Virginia wild rye, Illinois bundleflower and other native grasses — responded, announcing their arrival in areas where they had not existed before. By constructing a "waste area," McGrady in effect told the new arrivals it was safe to come out.

A cow should not have to travel more than a quarter mile for water, certainly not over half a mile. Watering stations are easy to construct with modern pipes and materials.

Weeds, much like water, have a vested reason for being. Many have medicinal qualities, and cows know to select those that qualify. They select the correct ones up front, even before they seek grass. Weed taproots bring needed minerals to the surface the way a dandelion recovers calcium that has migrated well down under.

A paddock isn't a golf course, but after it is operating, it is clear — poisons do not belong.

12

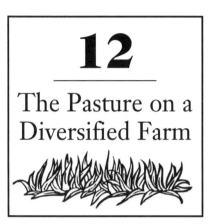

The Pasture on a Diversified Farm

One thing all consultants learn soon enough is that they have to make the scene. Installing fence lines according to diagrams conjured up on a drawing board is helpful only when the ground has been walked, analyzed on-scene, each break, tree line, open land and forest land. Questions that rate attention in Virginia might not even surface in western Kansas.

Joel Salatin, a diversified farmer in the Shenandoah Valley of Virginia, has refined forage paddocks in terms of multiple grazing. Willow leaves are excellent forage because of their medicinal qualities, for which reason a creeping crop just off a watering area might be tolerated, whereas less palatable species at the edge of the woods ask for regular trimming. Decision and indecision attend the workaday business of stringing fence wire. Salatin runs his electric fence wires along all the naturally defined edges to protect the riparian and forest zones at typical break points. Fields and forests, slopes and ridges as defined by direction in effect construct a map of the farm. The objective is to create homogeneity in the paddock. There is a reason for this special attention to terrain. A south slope may tend to be dry, but it greens up earliest in spring. A north slope holds its moisture more deliberately, but it lags in furnishing green forage as winter fades away. Obviously, a north and south slope in the same paddock might lead to overgrazing, undergrazing, premature grazing, even late grazing, all in the same poorly designed paddock.

The key word is flexibility. A swamp area might be generally off-limits, and its forage might be demanded when a dry cycle arrives perhaps once or twice a decade. Such an area suggests use only as needed.

The purpose for holding woodland off-limits in the Salatin scheme is to grow wood chips and carbon when grazing is not acceptable. When the soil is dormant and grazing is not acceptable, carbon comes to the fore. It's in demand for poultry in the hoop houses, cows in the hay shed, for pigs in vacated rooting areas.

Other farmers in other climes graze the forest. But this Virginia farmer draws sunlight via the agency of photosynthesis for his animals just the same, this by transporting carbon into the open land.

Woods have a unique function. Mycorrhizae thrive in the shade, and a carbon-rich soil builds and holds in escrow the very carbon that open swards demand. Some cattle growers tap forest-floor soil for direct feeding to cattle. Cows yearn for this carbon, and some root for it. As a compost starter, a live forest floor is nonpareil.

Forage obeys the rhythm of the season. As a consequence, bulk production seasonably outpaces nutritional needs of the herd or grazing animal population. Salatin calls hay "deferred grazing" — "We're taking that growth spike in May and deferring it to the dormant season."

A basic objective of the Salatin handling equation is to keep urine and animal manures off the soil when it is freezing. "The soil is the earth's stomach," the Shenandoah Valley farmer says. "It is alive and well." In fact, the soil microorganisms go dormant when frost penetrates the soil. Composters know this and all but the most skilled defer composting chores when the temperature drops below 50 F. In process, compost heat will sustain a windrow or pile. But a compost procedure traded off a below freezing temperature will likely flicker and go out, if such a metaphor can be allowed.

The art of building in suspension the 50-pound mother lode of manure with tree carbon pending the arrival of spring is as much a part of forage management as regulating the grazing term of the grass itself.

That forage management is no armchair job became evident to Salatin some two decades ago. Observation told him that feeding in the open was counterproductive. Not only were feedstuffs wasted, the stomping and urine mixed feeds are often eaten, with resultant infection being passed along replete with a full measure of debilitation. Fed in the hay shed, manure mixed with carbon, a measure of predigestion of manure takes place. Pests get their comeuppance in the wintery pile, this from the cold of a January spread. The identical manure fix is spread around at the beginning of March. By April the first spreading has disappeared. The March spreading atop the earlier distribution produces a lush stand of forage, dark green and loaded with minerals. Of all plants on planet Earth, grass is the grand champion in uptake of minerals.

The key to sanitation is feeding hay from a bulk feeder and maintaining a carbon-nitrogen ratio of 30 to 1. Over 30 to 1 destroys the compost, under 30 to 1 means smells take over because there is not enough carbon to suck up the nutrients. As manure is deposited by the animal, the ratio is 18 to 1. Sawdust clocks in at 500 to 1, wood chips are about 20 to 1, straw is about 60 to 1, and junk hay is about 35 to 1.

There is more art than science to forage management. When cows are out in the field in January and February, they do a lot of foraging damage. This takes five or six years to correct.

Cowmen scoff at the Salatin system and only a few deign to try it. Once the cows leave their winter feeding quarters, Salatin brings on his "pigaerators." The pigaerators go through the compost "like big eggbeaters," in effect oxygenating the anaerobic material, converting it from anaerobic to aerobic. Pigs love this chore at least as much as they enjoy rooting in a damp forest floor.

This preservation of the economic value of manure is seldom computed by bean counters, but the values are real just the same. Instead of expensive machines doing the work, pigs earn the value of their keep. Is it realistic to compute a value of 50 cents a day for the value of each cow's manure? Bookkeepers with an eco-farming background would say so.

The business of using pigs to aerate carbon-laced manure led Salatin to the realization that pigs like cleanliness, especially the cleanliness that attends grazing. This back-to-basics fellow views turning machines, windrows and picture-book refinement as Neanderthal stuff, especially when animals are fairly aching to go on breaking records delivering value to the bottom line.

It's all a point of view. An animal compost-turner might well be considered an appreciating rather than a depreciating piece of equipment. In terms of accounting, the profit potential becomes size-neutral simply because it is no longer a requirement for growers to capitalize all the green or orange equipment the clanging mart has to offer, all of it to be depreciated. Even a small farm can survive if it buys its "tractor" for $30 and sells it for $400 after its work is done.

Pigs make short work of their composting chore. This leaves some four months of work-free tractor-growing time.

Grazing pigs is an art for which the time has come. Salatin explains, "The pigs have a low center of gravity. They avoid the electric fence and enter their specially prepared converted forest. The trees have been chipped. Branches have been converted to chips. Firewood has been saved. Saw logs have been milled. Pig pastures have been divided into quarter-acre lots."

Each species inspects the land as they graze it. The long-term goal is a sustainable savannah. Merely cutting the timber won't produce pasture. The succeeding growth on the Virginia landscape would be brambles, brush, blueberries, not forage. Pigs impact soil the way they close up the lack in a farm. This is a negative for long-term gain. It has been reckoned that some 40 species of seeds per square yard shower down on this future pasture, origin the environment. Here art comes into play. Left too long, the soil becomes so compacted that nothing will grow, and erosion will be the legacy. Not left long enough, brambles and briars reclaim the area. In quarter-acre paddocks, a single ton of supplemental grain works perfectly as a measuring stick for the paddock moving to forage.

Pigs are omnivores — they can eat shrubs, forage, grain, even alfalfa hay.

Free-range swine do not comply with the much-advertised "other white meat" of TV fame. Grazier and all-around diversified farmer Joel Salatin hoots at the slogan. "They ought to say *the other white, slovenly, lack-of-nutrition, no-taste, cardboard junk meat!*"

When pork is raised in fresh air and sunshine with green forage, the meat has a deep rose color and has a greater density. The texture is different, and the taste is out of this world.

These rocky Appalachian pastures are rotated only three times a year, Salatin said. A six or seven time rotation, as with cattle, and the land goes backwards. Used as directed, the three pig pastures described above net $13,000 an acre. The count: three pig pastures, two acres each, eight quarter-acre paddocks in between, two strands of electric fence, one 4 inches off the ground, a second 12 inches off the ground, the electricity supplied by solar fence chargers.

Feeding supplemental grain seemingly does not affect the desired omega-3/omega-6 ratio. The reason is forage.

All animals, even dogs, eat forage. Even carnivores like a salad now and then. There is an aside to this comment on the vegetable question: Once vegetarians learn that the landscape can be healed with animals, it helps remove the guilt from eating animals.

In a world that sees bovines turned into feedlot ruminant hogs, it needs iteration and reiteration that herbivores are really a 100 percent forage animal. There is no defensible reason to feed an herbivore any grain whatsoever. Physiologically, the ruminant is not set up to handle grain, which is why *E. coli* migrates out of the colon, why acidosis calls in truckloads of Arm & Hammer bicarbonate of soda at feedlots, and why it is hardly possible to have a satisfactory dining experience with commodity meat.

Birds are the real grain feeders — turkeys, 50 percent grain; broilers, 15 to 20 percent grain; layers, 20 to 25 percent grain. Goats are browsers. They want to eat above their shoulders, not below.

The anatomy of plants requires meditation and study. Most weeds and all shrubs and trees require three times the rest period necessary for grass to recover growth. Indigenous peoples in the Americas seemed to know this. If lightning didn't fire the prairies, the Indians did. Fire would set back the brush and allow grass to claim the area. However, brush not fired down claimed the territory, as one often observes in the Hill Country of west Texas.

Lessons such as this one tend to expand themselves. Intensive grazing, for instance, holds down the slower-growing herbaceous plants and allows grass to rule the kingdom.

Goats are browsers. This means their rotation of habitat calls for innovation, space, time — otherwise these animals eat themselves out of a habitat. Once they do they start grazing, and with grazing in a goat pen goes a fix of parasites. Gastrointestinal problems follow. Nomads who didn't permit destruction of habitat usually worked their goats in a 40-year cycle. Anything less saw the destruction and desertification so commonly associated with the Middle East.

The classic pattern developed by sustainable goat cultures kept one-third of the land for sheep, one-third for goats, one-third for cattle. Cows, always, were the grazers.

A new area invited goats for 10 to 15 years. They cleared up the brush and allowed grass and forage the time required to endure. Sheep, with a higher nutritional requirement, then grazed 10 to 15 years. With a wider palatability index than the cow, sheep cleared up the less succulent weeds and kept the grass in trim. Finally, the cow harvested the land for 10 to 15 years. Unfortunately, cows avoided brambles and various brush plants, trees included. Thus the time arrived to bring on the goats. Some tribes ran cattle, goats and sheep together in a complementary grazing mode.

It takes an on-scene observer to detect nature's secrets. For instance, the pig does not do well on forage until it is 200 pounds.

Reading the above, it becomes evident that the industrial model for agriculture appears to be at war with nature. The quest for rapid return on investment has eliminated the component of the animal's life in which it performs well on pasture. This penchant has encompassed every aspect of husbandry. Breeding, selection, infrastructure, college research, genetic engineering, all have pursued the unattainable goal of having animals thrive on starvation and deliver quality out of the hell of confinement feeding, birthing and growth.

It takes a diversified farmer to compute the real values and destroy the logic of the industrial model. When cows were grass fed and grains still had to be harvested by hand, chicken was the luxury meat. It was the mechanization of grain production, funding world prices for grain with subsidies paid to large producers, that in effect subsidized cheap poultry and also feedlot beef. Even the hog demanded by the market dropped to 240 pounds once farmers were trained to cease grazing hogs to, say, 400 pounds. Today hogs are factory-grown, tasteless meat, all of it subsidized via grain subsidies to corn producers. Grain at parity would close down feedlots and confinement hog production almost overnight, and cows fed out on pasture would earn a better return for the cow-calf operator and rancher than the quick trip to the feedlot and the even shorter stay in an environment of fecal dust and tortured existence.

"As soon as farmers realize the benefit of converting row-crop land back to pastures, it'll take $10 corn to get farmers to rip their land up again. As soon as corn goes to $10 a bushel, the luxury meat again will be chicken," Salatin opined.

The animal on forage makes possible stimulation of omega-3 and omega-6 fatty acids, it colors up the meat and installs more iron, more riboflavin, all the B vitamins, the carotenes, everything that makes for a better cut of meat protein. Also, more size on an animal allows a better stand of amino acids.

There is one rule of nature that needs to be engraved above every college of agriculture. It simply states, "Herbivores never want grain."

An added inscription might point out that grass is more than the forgiveness of nature. It is the prime collector of solar revenue. Using the green coloring dye called chlorophyll and the agency of photosynthesis, grass collects solar energy, combines it with water and a few earth minerals. Animals do the harvesting of this energy as they turn it into meat protein. On the farm described here, even the fence is solar powered. Thus, nature becomes the template. In so doing, the grass farmer avoids rather than creates problems that cow colleges seek to solve.

Cows do not put milk into the udder while they're grazing, either free roaming or as a mob. They perform their metabolic chore while resting. Cows will not graze more than eight hours a day, thus the grazier's goal is to have animals ingest the greatest volume possible in the least time, then lounge.

Every species has a different metabolism. The cow holds up to 20 percent of her body weight in her digestive apparatus. The chicken, on the other hand, is not an efficient grazer. It can only hold 2 to 3 percent of its body weight, thus its reputation as an eating machine.

It takes 30,000 to 40,000 pounds moving over the land once or twice a year to push forage forward. Eighty cows on half an acre a day puts 160,000 pounds on the other side of the equals sign. A brief visit and rest during the remainder of the year — this is using animals as a landscaping tool.

Pastures on the diversified farm have to accommodate all grazing animals, chickens included. Chickens do not like long grass, for which reason sheep come to the rescue. They graze ahead of chickens that cooping systems disgorge on signal by the farm management.

The Salatin operation referred to above has choreographed a symbiotic relationship between rabbits above, chickens underneath along with pigaerators — all so the pasture can rest during winter months.

Pigaerators root up their pen so chickens can have easy access to worms up from deep hiding. Worms, of course, cycle what rabbits drop.

Withal, it is the stacking of operations that enlarges income, whether the source is permanent pasture, reclaimed woodland or row-crop acres returned to grass.

The bottom line: stacking, symbiosis, spreading the animals out, giving animals the air they need — 25 cubic feet per minute in the case of the cow! Open husbandry puts a dead end for pathogens into the equation.

Young chicks on grass via rolling or sliding pens — each day a new salad bar, each day droppings left behind as fertilizer — is rapidly becoming a part of the forage package as eco-farming reaches for a new tomorrow. In the end, grass — nature's own benediction — stays on to harvest still more solar revenue.

13

A Sea Energy Pasture

When Lewis and Clark crossed the American continent on behalf of Thomas Jefferson, the virgin country was forest and prairie. The bison roamed freely in vast herds all the way from Great Slave Lake in Canada to northern New Mexico, as far east as Pennsylvania and west to the valleys of the Rockies. Great Savannas of native grasses governed areas so vast they produced hardly a tree, crowded out shrubs, and purified themselves with fire from lightning strikes and Indian assistance.

The buffalo did not overgraze. The herds ate their grass, deposited their fertility, and moved on, allowing the grass to recover. As the late William A. Albrecht often pointed out, the bison identified the best soil, the best forage. This animal ate only grass, and it selected its grass with the skill of a connoisseur.

Although hunted to near-extinction, the bison survived and is now grown in captivity by some few ranchers. The term "captivity" does not imply restraint. Clearly, the buffalo is restrained only voluntarily.

Frank Polifka of Catherine, Kansas, once tried to restrain a buffalo herd in a regular bovine squeeze chute. The animal simply demolished the chute.

The buffalo wants grass, grass before it has jointed, grass loaded with nutrients. It shuns rubbernecks with cameras and simply refuses alfalfa hay.

When western Nebraska rancher Don Jansen was spraying pastures for weeds, using the best poisons LD_{50} identification had to offer, his tank ran dry with some few acres left to go. Jansen had some ocean crystals harvested from the Baja Peninsula of Mexico by the physician Maynard Murray. The importation of these materials into western Nebraska had become a reason for coffee shop merriment. It was common knowledge that the egghead professor was ruining the land his father and brother had assembled into a 15,000-acre ranch before and after the Korean War.

"When no one was looking, I put one coffee can full of the crystals into the tank with water and finished spraying the field," Jansen later recalled. He was a bit reluctant to admit that he had become a rebel convert to ocean mineral fertilization.

One of those rare dryland rains intervened, and within days the specially sprayed pasture greened up. The buffalo, usually grazing as far from road traffic as they could get, suddenly forgot their shyness. Clearly, they relished the new readily endowed forage Jansen came to call sea energy grass.

"They grazed that grass with the precision of a lawn mower," Jansen said. "They selected the trialed area for resting as if the ground now had a new energy to offer."

The incident was what college researchers call anecdotal evidence. It was not to remain anecdotal for long. Growers in and around the midlands heard about sea energy agriculture. In fact *Sea Energy Agriculture* became the title of a short book by Maynard Murray and Tom Valentine, which detailed the experiments of Murray, an eye, ear and nose doctor who became interested in the potential of ocean solids as a fertilizer while on travels that took him to sea for nine months. Murray could hardly help noticing that fish and whales enjoyed pristine health while relatives in rivers and estuaries suffered a cancer profile not much different than the one troubling mankind. He performed necropsy examinations of 80- and 90-year-old whales, and compared their internal organs to those of babies. The lab reports came back. The examiners couldn't tell the difference.

Murray and his former associate Ed Heine ran batteries of tests on corn, various commodities, vegetables and fruit trees. When a flash flood carried some of their sea solids into the lower reaches of a neighbor's pastures, they noted the cattle preference for grazing in an area previously shunned.

Toxic Agriculture

Shortly after the Poison Control Centers were established in 1949, American agriculture adopted the toxic norm. Even the nature-loving farmers bowed to their intellectual advisers and instilled all the apparatus and all the poison that industrial agriculture has to offer. Don Jansen, then a young man, rebelled. Even before he was old enough to vote, he'd had his fill of dark-to-dark work. He hitch-hiked east, worked his way through several colleges, and became a duly tenured professor. "I had it made," he summarized — and then tragedy struck. His brother, tormented beyond endurance by tumors, clearly a consequence of the technology now a part of farm life, ended his life. Dad Jansen was too old and too sick to run such a large operation.

Don Jansen had grown up on a farm, but farming had become an insane exercise based on warehouses full of packages that smiled at one and all with skull-and-crossbone warnings.

During this difficult period, while going from hospital to hospital, doctor to doctor, with his father, Jansen encountered *Sea Energy Agriculture.*

Here was an abundant supply of nutrition, the 92 non-radioactive elements in the ocean itself. The physician Maynard Murray had identified the sodium balance level of soils. It complied with the cation balances that Missouri University under William A. Albrecht had offered as settled science. Jansen ordered up a truckload of sea solids, this after selling off poisons and some of the machines. He sprayed crops and he sprayed pasture.

During the 1960s, public policy decreed that there were too many farmers, that some were not efficient, that agriculture had to feed the world, and that this could be done only when the land was in a few strong hands. These few pronouncements became current coin even while farms and ranches had become too big for management without the use of toxic genetic chemicals. Jansen had no intention of becoming yet another sacrifice on the altar so slyly erected in the devil's pantry.

He sold out, retired to Florida, bought Maynard Murray's Seaponic Farm, and reflected on the lessons a herd of bison had imparted and which no college seemed to teach.

Before the plow turned the prairie upside down, as the Plains Indians put it, nature developed grasses and saw to their nutritional needs. Oceans once covered all the lands, even the heights of Everest, and glaciers more recently scattered a mix of nutrients across the continent. The ocean remains, all the nutrients in a perfect mix, excesses canceled out by forces as difficult to comprehend as unlimited space itself. Utilization of this resource would appear devoutly to be wished.

Possibly the first to this one part of the Creator's plan was René Quinton, a French biologist who lived well into the 20th century. He discovered that human blood was the same as primordial ocean water. We came from the sea and we reach for the stars, but it is the sea, not the stars, that best shape our destiny. Quinton used ocean water as intravenous therapy with great success, but he never thought to use ocean minerals on pasture. Not even André Voisin came to this appropriate conclusion, even though his *Soil Grass & Cancer* is a paean to the dusts of the ocean's essential minerals in trace form.

Grass grows, on poverty soil, on rich silt loan. Grass on poverty soil — in Florida, Mississippi, in the rainbelt east of the Isohyet can be enriched with ocean energy, as some few graziers have proved while neighbors saw their cattle incapacitated by hip-high but deficient grass.

The Treatment

The treatment for ocean-grown pastures is simple in the extreme. Ocean solids or ocean water has to be diluted and sprayed, preferably in the evening. The dilution depends on whether the ocean water is concentrated, or whether the crystals have been harvested from one of several uncontaminated beds located at strategic points on plant Earth. Physician Maynard Murray identified those beds while on his ocean odyssey.

Agronomy has no real answers or explanations with reference to ocean-energy techniques. Indeed, the roles of most trace nutrients have still to be defined. This usually happens the day someone learns how to package them and make them commercially available, at which point they become essential.

It is the postulation of Jansen — and before him, Maynard Murray — that *all* the elements are essential for maximum health.

Few vegetables and no natural grain crops are able to assemble as a single nutritional package the minerals and vitamins available in fresh unjointed grass. Tomatoes have checked in with a package of 56 trace nutrients, sweet potatoes with 70, and other crops with like diminished numbers. Only grass seems capable of assembling 92 elements, and this only in the annual. Just the same, perennials do a mighty job.

The responses to ocean dust have been dramatic ever since the physician Maynard Murray first counseled Illinois, Indiana, Iowa and Wisconsin farmers to use ocean solids on pasture forage. The shortfall of trace nutrients cannot be over emphasized. Wendell Owens, the good right arm of Dan Skow's International Ag Labs in Fairmont, Minnesota, tells me hardly anyone tests pastures for cobalt, and yet the absence of that trace mineral helps usher in substantial health problems for the herd.

A major problem in almost all perennial pastures is compaction. Often the wrong major grass is in place, and penetrating roots are missing. More likely, too many traces are missing. Accordingly, the tiny livestock in the soil fails to keep it open so that it can inhale and exhale with the swing of the moon and the shuffle between night and day. Mechanical aeration is an option.

The best graziers test routinely for copper, iron, manganese and zinc. "Copper is nature's fungicide," noted Owens. "In wet years you can get fungus in the grass. Copper keeps that out." Probably 90 percent of the pastures are deficient in copper. Too many goat, sheep and horse farmers depend entirely on supplementation instead of dealing with the grass and its hungers.

Until quite recently, these traces were a best-kept secret. The ocean source utilized by old man Murray was ignored, and Jansen's findings seem to be almost alone in recapturing the values researched a half-century ago. Almost everyone knows the rule-of-thumb stuff. Manganese makes the grass go to seed, but when an excess develops, it can inhibit the uptake of copper. The 90 to 92 traces Jansen's associates stress seem to take the poverty out of poverty grasses.

Table 1: Average Percent of Elements in Earth's Crust (10-Mile Shell)

Oxygen (O)	46.6	Copper (Cu)	0.007
Silicon (Si)	7.7	Tungsten (W)	0.0069
Aluminum (Al)	8.13	Lithium (Li)	0.0065
Iron (Fe)	5.0	Nitrogen (N)	0.0046
Calcium (Ca)	3.63	Cerium (Ce)	0.0046
Sodium (Na)	2.83	Tin (Sn)	0.004
Potassium (K)	2.59	Yttrium (Y)	0.0028
Magnesium (Mg)	2.09	Neodymium (Nd)	0.0024
Titanium (Ti)	0.44	Niobium (Nb)	0.0024
Hydrogen (H)	0.14	Cobalt (Co)	0.0023
Phosphorus (P)	0.118	Lanthanum (La)	0.0018
Manganese (Mn)	0.1	Lead (Pb)	0.0016
Sulfur (S)	0.052	Gallium (Ga)	0.0015
Carbon (C)	0.032	Molybdenum (Mo)	0.0015
Chlorine (Cl)	0.031	Thorium (Th)	0.0012
Rubidium (Rb)	0.031	Germanium (Ge)	0.0007
Fluorine (F)	0.03	Samarium (Sm)	0.00065
Silicon (Si)	0.03	Gadolinium (Gd)	0.00064
Barium (Ba)	0.025	Beryllium (Be)	0.0006
Zirconium (Zr)	0.022	Praseodymium (Pr)	0.00055
Chromium (Cr)	0.02	Scandium (Sc)	0.0005
Vanadium (V)	0.015	Arsenic (As)	0.0005
Actinium (Ac)	0.013	Hafnium (Hf)	0.00045
Nickel (Ni)	0.008	Dysprosium (Dy)	0.00045

Uranium (U)	0.0004	Cadmium (Cd)	0.00001
Boron (B)	0.0003	Silver (Ag)	0.00001
Ytterbium (Yb)	0.00027	Indium (In)	0.00001
Erbium (Er)	0.00025	Selenium (Se)	0.000009
Tantalum (Ta)	0.00021	Argon (Ar)	0.000004
Bromine (Br)	0.00016	Palladium (Pd)	0.0000017
Holmium (Ho)	0.00012	Platinum (Pt)	0.0000005
Europium (Eu)	0.00011	Gold (Au)	0.0000005
Antimony (Sb)	0.0001	Helium (He)	0.0000003
Terbium (Tb)	0.00009	Tellurium (Te)	0.0000002
Lutetium (Lu)	0.00008	Rhodium (Rh)	0.0000001
Titanium (Ti)	0.00006	Rhenium (Re)	0.0000001
Mercury (Hg)	0.00005	Iridium (Ir)	0.0000001
Iodine (I)	0.00003	Osmium (Os)	0.0000001
Beryllium (Be)	0.00002	Ruthenium (Ru)	0.0000001
Thulium (Tm)	0.00002		

SOURCE: Chemistry, *by Gerard F. Judd, Ph.D.*

The pasture enzyme has been confirmed through the years, a late entry arising from the University of Missouri. The results of a five-year study reveal that half the cows in the state were mineral deficient, especially in copper and selenium. The study did not consider the full range of nutrients found in ocean water. The Missouri study involved approximately half the counties in the state.

Fully 30 percent of the cattle tested exhibited some level of mineral deficiency. The several shortfalls noted routinely account for reproduction problems and growth and general health. Grazing practices were indicted in terms of causality when pinkeye occurred, a consequence of close grazing, a Texas study some few years back — and which was quoted by Missouri researchers —

revealed that close-cropping often deprives cattle of nutrients. Pasture improvement should rank at least as high as supplementation.

Carbonatite

Over the past 35 years, new product lines have arrived on the market. For most practical purposes they have literally annihilated the "settled" science of N, P and K.

Their application for pasture maintenance is the business of the promoters, who must take on the responsibility of making good claims.

Some of the dusts are available at quarries and mines, strip and deep shaft. Others are being commercialized, often as innovations, always without a structural distribution system.

I cannot cover them all. Malcolm Beck has assembled hundreds of samples with paramagnetic properties. Harvey Lisle wrote a charming little book a few years back that merits attention. It is titled *The Enlivened Rock Powders.*

The latest entry to arrive surfaced in Canada as carbonatite.

Biotite does away with the idea that potassium should be soluble. As William A. Albrecht put it, the nutrient should be insoluble but available.

An unlooked-for consequence of the eco-farming movement has been a side departure for researchers into the enigmatic world of organiculture studies dealing with the prairie states and provinces of Canada.

High-energy rock powders serve up a reaction in the same year as applied. It is up to the pasture to provide a balance of minerals, vitamins, enzymes, protein and energy.

Minerals are the foundation for life, even if many have still to be researched and understood. During the last 35 years new attention has been directed at traces. These production-limiting factors in the pasture can be remedied from every direction. The first big step is to consider forages as more important than the highly touted cereal crops.

It is an interesting exercise to compare the table of minerals available in ocean water with a similar table of minerals in the top 20 miles of earth. The comparison can be seen in the tables reproduced in this chapter.

Table 2: Major Elements in Ocean Water; percent & Million Tons of Element per Cubic Mile of Ocean Water (4.75 x 109 tons)

Elements	**Percent**	**Million tons, amount per cubic mile of ocean water**
Cl (chlorine)	1.9	90.3
Na (sodium)	1.05	49.9
Mg (magnesium)	0.135	6.40
S (sulfur)	0.0885	4.2
Ca (calcium)	0.04	1.90
K (potassium)	0.038	1.81
Br (bromine)	0.0065	0.31
C (carbon)	0.0028	0.13
Sr (strontium)	0.0013	0.062
B (boron)	0.00046	0.022
Si (silicon)	<0.0004	0.019
F (fluorine)	0.00014	0.0067
Al (aluminum)	0.00005	0.002
Li (lithium)	0.00001	0.0005

Million tons: H_2O 4590 salt 155

SOURCE: Chemistry, *by Gerard F. Judd, Ph.D.*

Table 3: Composition of the Atmosphere

Elements	Parts per million at STP (by weight for soilds & by volume for gases)
O (oxygen)	0.0781 x 106
N (nitrogen)	0.2095 x 106
Ar (argon)	9340
Ne (neon)	18.18
He (helium)	5.34
Kr (krypton)	1.14
O (ozone, O_3)	0.0 - 0.047
H (hydrogen, excluding H_2O)	0.5
Xe (xenon)	0.087
S (sulfur)	0.001 - 17.5

Compounds	Ppm (parts per million)
CO_2 (carbon dioxide)	330
H_2O	3.7 - 3700
N_2O (nitrous oxide)	0.26 - 0.61
CH_4 (methane)	1.2 - 1.5
SO_2 (sulfur dioxide)	0.0 - 0.035
NH_3 (ammonia)	0.0 - 0.02
H_2S (hydrogen sulfide)	0.002 - 0.03
NO_2 (nitrogen dioxide)	0.0 - 0.0029
CO (carbon monoxide rural)	0.0008 - 0.016
CO (carbon monoxide urban)	0.0008 - 0.016

SOURCE: Chemistry, *by Gerard F. Judd, Ph.D.*

Some few local deposits lay claim to a full-spectrum inventory — "full spectrum" being taken to mean double or even triple the requirements acknowledged by settled science. Azomite is a commercially available mineral deposit that alleges the usually available roster. A newfound deposit from Canada is proving itself in pasture management because it invites comparison with ocean water in mineral content.

Carbonatite is coughed up by volcanic activity from the magma crust and rates attention as a rare calcium source from an igneous rock source.

The story of how fertility from the center of the Earth is pumped into the ocean via vent and volcanoes is told in *Fertility from the Ocean Deep*. Equally fascinating is the content of trace nutrients (in ocean amounts) from magma to the surface, giving surface dwellers calcium from an igneous rock source, using geologist John Slack's words. This source, mentioned above, is called carbonatite — calcium laced with traces. It shatters when it comes in contact with air. It sizes itself in the ion and angstrom neighborhood. It enters the ag stream as a pasture and soil amendment, repairing the trace nutrient shortfall that has become endemic.

Unfortunately, carbonatite represents only 1 percent of the minerals of planet Earth. It and its cousin kimberlite are very rare.

Hungry pastures want minerals, whatever the source, a point graziers are obligated to study, remember and implement.

14

The Brunetti Way

David Zartman, the pasture specialist at Ohio State University, makes a point of referring to pasture grass as *forage*. He points out with disarming finality that it takes more than blue stem *(Andropogon scoparius)*, Texas wintergrass *(Stipa leucotricha)*, and a raft of grasses with or without the Latinate handles to make a pasture. Almost everyone has a favorite. Extension folk can provide genus and species, and they will provide suggestions according to the area of the country.

Jerry Brunetti of Agri-Dynamics has appeared at seminars around the country for years discussing everything from crop production, animal husbandry, veterinary medicine, human health, and, not least, pasture management. His taped message provides a rich, ripe, even racy catalog for eco-agriculture, the fountainhead for this insight being pasture management. And pasture management starts with diversity. And diversity starts with the proposition that pastures left ungrazed are no better off than pastures grazed too much, animal manure being an ecological necessity.

Diversity in the pasture is a must because the very term implies the presence of medicinal plants. Nature seems to have decreed such integration as a requisite for ecologically correct poultry, egg, milk and meat protein production. Over the past 50 years, manure has left behind its time-honored role as a valuable resource and become a disposal problem. This migration of insight has walked hand in hand with the "get big or get out" syndrome that has attended the public policy of delivering land into a few strong hands.

That the mega-farm structure with its feedlots, confinement feeding, manure pits and lagoons is poor ecology, may be beside the point in the business equation. The disposal systems are poor economics. Manure pits that often become five million cesspools tend to define industrial model dairy farms. The term ecological disaster may well be the understatement of the hour.

The idea of hauling great volumes of manure with pits and lagoons was spawned by the factory in the field vision that relies on total mechanization.

Application of a lagoon product to topsoil with the intention of recovering value biologically choreographs the collision of two eco-systems, an anaerobic toxic ecosystem with an aerobic biological system. If anything, this fix worsens the soil system.

Consider the makeup of the product. It is laden with salt and sulfides. Sulfur is wasted away. Sulfur is difficult to hold in any case. It is an anion and tends to leach. Also lost in the lagoon — soil exchange is ammonia simply because nitrogen is reduced to ammonia.

All of the above translates into a wasteful, ecologically destructive way of dealing with manure.

Old-timers handled bulk manure via anaerobic fermentation. Anaerobic fermentation and pit refraction are not the same. Fermentation is the kitchen's sauerkraut system installed in the manure pile.

All this is an aside simply because the pastured animal continues to be ruminant and not the monogastric feedlot creature with the attendant problems of acidosis, *E. coli* migration, feedlot bloat and debilitation due to fecal dust forever assailing the lungs. On pasture, the bovine gets 25 cubic feet of fresh air, a commodity never available in close confinement.

The eco-grower has a vested interest in reducing the need for pharmaceutical drugs, not only because of side effects and reverse effects, but also because of the prohibitive cost. Moreover, informed consumers will no longer accept drug residues in milk, eggs and meat.

As noted in *Weeds: Control Without Poisons*, close to 50 percent of the weeds listed in weed manuals are also listed in medicinal botanical manuals. Apparently the bovine animal has had the instinct needed to select the right herbs and forbs for health maintenance for centuries, and science is only now picking up on this intelligence.

All available evidence suggests that stands of pure grass, whatever the species, is no medicine chest for the animal. In fact it is the mixture that contributes to gain in milk and meat protein, and not the pure stand of fescue or alfalfa or clover. Management of cold season cultivars and permission to live for indigenous herbs — including deep-rooted perennials — bless pastures with unmeasured nutrient density, drought resistance, palatability, always exhibiting perennial persistence and soil conditioning characteristics. They have a capacity for uptaking minerals. Indeed they are specialists in this role. They can tell the farmer what's wrong with the soil and what's right with the soil.

Almost all pasture specialists suggest forage mixtures, usually four or five varieties. Few reach for the grand diversity found on the high plains where climax grasses not only built soils, but also preserved them from the twin engines of erosion — wind and water.

Newman Turner, writing in *Fertility Pastures*, discussed the importance of roots reaching the subsoil, a consideration that makes the dandelion, which so offends the urban psyche — a most valued plant. Its roots tunnel down three feet or more, returning valued calcium to the surface. Turner suggested subsoiling every seven or eight years in order to penetrate an expected hardpan. Herb and forb penetration makes this exercise unnecessary. Turner cites chicory, burnet, lucerne and dandelion.

"I have seen my Jersey going around patches of nettles or docks eating off the flowering tops, relishing something they have been unable to obtain from the simple, shallow-rooted legume mixture," wrote Turner.

Turner's matchless masterpiece counseled getting back into the dryland pastures as many herbs as possible, this to give the animals a botanical pharmacy only animals know how to use to the fullest extent.

Burnet, cockerel, plantain, wild vetch, sheep's parsley, sweet clover, chickweed, all have a defined role to play. Once these denizens of the pasture survive to send down yard depth roots, they are most liked by cattle. Turner, who Jerry Brunetti quotes at the drop of a hat, often told of cows migrating from shallow rooted pasture to acres where the deep rooted plants performed their holy task, all because new and rare resources were being tapped from well below tillage levels.

As a consequence, "bloat has become a thing of the past," whereas on pastures seeded with only three or four forage varieties, Turner lost cattle every year.

Turner makes a point of recommending comfrey, garlic, raspberry, hazel nut, docks and cleavers, etc., all in hedge rows.

With plenty of dandelions grazing in grass swards, Turner found a lime profile always adequate, lime tends to migrate lower in the soil profile. The dandelion brings it back to the surface.

Turner's bottom line still stands unrefuted. "It is now evident that organic methods which include subsoiling and deep rooting herbs over a period of years maintain a correct soil balance, even on farms which are sending away large quantities of milk."

Jerry Brunetti, whose body of work is and will be quoted directly and indirectly, adds a codicil.

Annuals pastured late in the afternoon harvest non-structural carbohydrates in abundance. Dry matter intake goes up, and milk production goes up — 8 percent according to one USDA study, all a consequence of cutting hay or grazing annuals late in the afternoon. This aspect of grazing is often overlooked. The grazing animal encounters starches, sugars, and structural carbohydrates. In reality some of these fibers are quickly fermented, much later sugar. These are the pectins and beta glucose. Pectins dominate in legumes, beta glucoses dominate in grasses.

Buying starch off the farm is a bottom-line killer. It can be grown in the grasses, forbs, legumes, in the grand miscellany that makes up the forage pasture.

It is axiomatic to note that the only creatures designed to consume grain are birds. That is why many old timers turned grain into grass via the agency of sprouting, this during winter or when pastures faltered. Sugars are generally characterized as sucrose, fructose, glucose, lactose and dextrose.

As grains are sprouted, starch changes to sugar. Sugars do not produce ruminant health problems. Not like starches.

Sprouted grain achieves digestibility. It delivers enzymes that make grain feedstuffs digestible. Enzymes and grains are antagonists. The latter are loaded with enzyme inhibitors, and the former work mightily at the business of undoing grain's stronghold. That's why a 24-hour soaking nullifies phytates in grain.

The syllogism delivers its last conclusion. Forages have high solubility of protein.

A victim of cancer, Jerry Brunetti came to terms with the degenerative affect studying nature from the ground up. No one dares use the term cure. It now belongs to the government, which has defined it, and perhaps to the saints. If we ask the bovine animal in a language the animal understands, cure belongs to the pasture.

Those who have recorded the answer tell us exactly which plants are preferred by Jersey cattle. Newman Turner, mentioned earlier and often a backbone reference when Jerry Brunetti takes to the podium, planted 35 individual plots, each with a single ingredient — a half pound of the clover or grass seed. The answers came back soon enough. The star performer in the preference department was sheep's parsley, then plantain and chicory. Way down the scale were rye grasses, meadow fescue and hard fescue. After these came burnet, kidney vetch, sainfoin and alsike clover. Alfalfa and American sweet clover were rejected the way Fido rejects a plastic hamburger, this when any of the aforementioned plants were available. When it came to grasses, the preferred species were short rotation rye grasses, meadow fescue. The hard fescue was simply rejected. The other grasses were tolerated once the preferred plats ran out.

Turner couldn't answer whether soil fertility or even the breed of cow governed the choice. His experiments were made at a time when brucellosis was a national scourge, possibly because of a near-total absence of cobalt in American soils. Weeds are specialists. But they cannot uptake a mineral nutrient if it isn't there. And by the beginning of the last century the physician and investigator George H. Earp-Thomas could declare with certainty that cobalt was totally absent from New Jersey soils. The absence of other nutrients during the last half century has also been noted, for which reason special treatment for pastures and specialized rotational grazing is rapidly becoming the norm.

Turner did not expand on thoughts he might have had regarding carbon dioxide in the company of forage swards. He merely posed the right questions, some of which have still to be answered.

Brunetti's fondness for texts by old masters became evident when he crafted a report for *Acres U.S.A.* titled "The Benefits of Biodiverse Forage" (October 2003), an abstract of which follows. He refers to research in the late 1890s as reported by one Robert Elliott in a classic book, *The Clifton Park System of Farming*. Cited were the properties of chicory and other forages purists shun the way the devil is said to shun holy water. There was a severe drought in 1895 in Scotland, the locus of Elliott's report. Chicory, burnet, kidneyvetch, and yarrow survived the drought.

Chicory was a pasture treasure at least as early as 1787, being cultivated in England that year by a man named Arthur Young. He had imported the plant from Italy, where it was as common as an urban dandelion patch. The plant proved to be absolutely prolific. It weighed in with 11 tons of hay per day compared to lucerne at five tons a day, six cuttings yielding 30 green tons in 1788, apparently off smaller plots. Elliott left us the legacy of his findings as data that Morrison's *Feeds and Feeding* might well have quoted. Elliott observed the roots of chicory migrating 22 inches in five months, and 30 inches in 15 months. When Thomas Jefferson heard of this remarkable plant record, and its propensity for growing in a miscellany of soils, he violated his own political push for an embargo on British imports and brought the plant to Monticello.

Simply stated, the plant tunneled well into the subsoil, took up nutrients shallow-rooted plants could never access, and conferred a cafeteria of nutrients on cattle, horses and hogs generally absent in grasses and legumes with shallow root systems.

As this information rose to the surface, traveling through the information capillaries of time, a mix was developed by the middle of century 20. It was a Turner mix, and it is still stressed by Brunetti as follows:

Chicory, lucerne, New Zealand rye grass, coxfoot, timothy, meadow fescue, perennial rye grass, late flowering red clover, S-100 white clover, sheep's parsley, yarrow, tall fescue.

This idea of herbal mixtures gathered speed as this investigator warmed to his subject.

An early gray herbal mixture, it cancels out the yearning for forced growth with nitrogen fertilizer and the attendant impact of energy reduction and production of non-protein nitrogen.

Mid-summer herbal ley. This was designed to endure drought damage.

Herbal mixture for autumn and winter grazing. This mix was designed to grow in late autumn and winter.

Herbal mixture for fragile dry soils. Here the deepest rooting plants available were conscripted for pasture preservation service.

All purpose ley mixture. This general mix was designed for general all-year service.

Herbal hedge growth mixture. As a good browsing and grazing mixture, this mixture was to be sown in or near hedge rows.

Turner's book is so treasured, few owners allow others to read, much less borrow it.

Of note is the role chicory has in pig and poultry leys, with added emphasis on plantain and burnet, and lesser amounts of sheep's parlsey, yarrow and kidneyvetch.

As we walk through the years, it becomes evident that all researchers stand on the shoulders of giants. Turner relied on and was inspired by Elliott, and Elliott no doubt tapped the oral wisdom of generations of graziers that developed the excellent breeds extant in the British Isles. You read lore that is all but forgotten by modern cowmen. Here are two fields with similar soils albeit seeded with different mixtures.

Field 1. Coxfoot, perennial rye grass, late flowering red clover, S-100 white clover, and one pound per acre chicory, a total of 25 pounds was sown for the acre.

Field 2. The same legumes and grasses as used in Field 1 were sown albeit with the following additions: 3 pounds per acre chicory, 4 pounds burnet, 2 pounds sheep's parsley, 2 pounds kidney vetch, 1 pound yarrow, 2 pounds lucerne, 2 pounds American sweet clover, a total of 45 pounds seed.

Establishment for these two fields was quite equal. When cows were allowed to pasture the second field, milk always increased.

Noted Brunetti: "There's more to nutrition than the usual parameters surrounding protein, energy, total digestible nutrients, neutral detergent fiber, acid detergent fiber, and so on. Perhaps the diversity of such a mixture in a paddock provides critical trace elements required for various plant harmonies, enzymes, aromatic oils, tannins, amino acids, fatty acids, alkaloids, pigments, vitamins, and co-factors, unidentified rumen flora, etc."

The humiliating fact is that animals used this nutritional knowledge long before expensive labs with credentialed scientists even gained an inkling of this knowledge. It is only when animals are ruthlessly confined that they consume the proffered rations conjured up by tenured schoolmen far away from the reality of fertility pastures.

When the pasture provides appropriate nutrition, veterinary bills — except for trauma — virtually disappear. Decades of experience taught Brunetti that animal health is a primary, not a secondary consideration when it comes to achieving the bottom line. This single consideration commands diversity in the pasture.

Ranchers are justly concerned about palatability and toxicity when strange plants invade their grassland acres.

In cooperation with Utah State University, the National Resources Conservation Service & Grazing Knowledge Institute and Utah Agricultural Research Station, researcher Fred Provenza compiled a vast amount of data on the topic. This is archived in a publication entitled *Foraging Behavior: Managing to Survive in a World of Change.* Provenza suggests that livestock develop a natural wisdom as a consequence of interaction between flavins, nutrients and toxins. Decreases in palatability happen when foods contain excessive levels of either nutrients or toxins. With feed causing nutrient imbalances and deficits, animals are able to discriminate between feed based on sensory feedback from nutrients, according to energy, protein and mineral levels. That is why grazing animals select a variety of pasture plants when they are available. It is an uncomfortable fact that all plants contain measurable amounts of toxins. These, however, are nature's molecules and do not equal man-made molecules such as 2,4-D, 2,4,5-T, and a thousand formulations many scientists believe have no safe level and no tolerance level.

Livestock know how to regulate their intake of nature's toxins. Moreover, many plants provide compounds that neutralize toxins "or activate metabolic pathways to eliminate them," the Utah researcher noted.

Mere recitation of these data and facts took consultant, researcher, grazing specialist and entrepreneur Jerry Brunetti to a conclusion he states with clarity.

"Since animals prefer traditional feeds to novel ones, rotational grazing methods that incorporate low stock densities may actually detrimentally modify behavior of genetics of livestock to eat the best and leave the rest, thus accelerating the decline in biodiversity."

Provenza suggests that heavy stocking for short periods encourages that mixing. Mother cows teach their young. This may start in the womb. The transition from milk to plants implies instinct, but it may also suggest a teaching mode.

Brunetti likes to sum up the findings of those on whose shoulders he stands along with his own observations. "Recognizing the fact that rhizospheres of plants are actually eco-systems in and of themselves, it is agronomically critical to take into consideration that the diverse number of species, perennial deep-rooted herbs, legumes, perennial grasses, annual grasses, biennial legumes and herbs, provide an indescribable substrate upon which a very complex food web can be established. The food web includes multiple species of bacteria, protozoa, fungi, arthropods, earthworms, nematodes, and so on. This diversity in the soil creates the same opportunity for the higher life forms that are dependent upon the plantation of the earth. These ecosystems are grasslands, rain forests, coral reefs or savannas. Life begets life continuously because predation, digestion and recycling occur effectively when there is this diversity."

While assembling the notes and papers upon which this book is based, Brunetti supplied a personal experience that literally asks for inclusion. He visited an example of monoculture myth in Ohio. There were two plots. One was seeded in perennial rye grass with glyphosate treatment. The soil had been fortified with lime, phosphate, potassium, boron and gypsum for sulfur. And nitrogen also a part of the treatment. The second plot had the same treatment sans nitrogen. Alsike clover, fescue, red clover and orchard grass were a part of the seeding.

The ryegrass-only plot "hit the ground running," as the saying has it, providing far more dry matter per acre. By mid-summer, the ryegrass plot exploded with an outbreak of rust. The diverse plot remained unscathed. Diversity seemed to be the insurance policy difference.

If this suggests a hydra, a creature with no single head upon which cause and effect can be hung, so it is. The interplay would overload the most powerful computer, the variables are that numerous.

The Miracle of Roots

The Elliott work cited above told of an experiment dealing with the penetration of a hardpan barrier. The soil was on a low lying alluvial flat. His mixture was accompanied by a thin seeding of oats, 5 pounds each of coxfoot, meadow foxtail and tall fescue, 7 pounds meadow fescue, 4 pounds timothy, 1 pound each of wood meadow grass and what he called rough stock meadow grass; 2 pounds each of white clover, alsike and perennial red clover, perennial vetch and lucerne, 3 pounds chicory, 8 pounds burnet, 1 pound chat parsley, and half pound of yarrow. The 15 acres were cut for hay, yielding 36 tons: 14 carts, 2.5 tons per load. Two trenches were cut through the middle of the field. The pan had been so tough, it took a powerful man to break it open with a tool. The trench revealed that chicory has disintegrated the pan. Roots of the best plants went down 30 inches. Burnet and vetch roots went down 20 inches. Lucerne made it 8 to 10 inches. Together, these plants did the work usually assigned to a subsoiler, this in a single year.

The conclusion appears to be obvious. In addition to the medicinal value of the diversity entertained by several investigators, that same diversity goes a considerable distance in healing the soil. Most likely, they make life for the subterranean workforce tolerable, for which reason they perform doing their thing, taking inorganic minerals into the plant, in effect giving plants a powerful assist at the business of making inorganic nutrients organic.

Now the pasture story enters the arena of the microorganism, compost and compost tea, even into the esoteric arena of biodynamics, which will be considered in the final pages of this volume. It takes a vibrant soil microorganism profile to handle phosphate and make it plant ready. The Haughley organiculture experiments have proved that composts made of diverse materials are best suited for transforming inorganic phosphates into organic phosphates.

Brunetti does not presume to dismantle the cation exchange system taught by the late William A. Albrecht of the University of Missouri. In terms of a base exchange of cations, the objective is to balance calcium, magnesium, sodium and potassium, as discussed in detail later in this chapter.

It is axiomatic that if the minerals are missing, the hope for excellent health and genetics is in vain. Even if the crude protein numbers comply with sanctified feed values. William A. Albrecht often noted that the only test that counts is the bio-test — feeding the grass, forage or feedstuffs *to* animals. How they thrive or fail outweighs lab numbers any day of the week, making production, reproduction and immunity to disease, health of offspring, meat and milk quality the only numbers that matter.

According to Brunetti, sulfur levels should be at least 10 percent of the nitrogen level. A 10 to 1 or lower nitrogen to sulfur ratio suggests that there is less non-protein nitrogen and therefore the protein has a more complete amino acid profile. Sulfur is also a vital component of the essential amino acid methionine as well as cysteine, precursors to glutathione a tripeptide anti-oxidant.

For those who are prepared to make the transition from grass and forage to what the medicinals in plants actually do, Brunetti unloads an inventory of knowledge that is dazzling in its purity.

For instance, methionine and cysteine are precursors to glutathione, a tripeptide, which also happens to be a building block of glutathione s-transferase, an important liver detoxifier. And glutathione peroxidase is a critical immune activator. Phosphate is a necessary element for energy molecules associated with the Krebs cycle. Magnesium is associated with over 300 enzymatic reactions, energy production in animals included. You can count the traces on the Mendeleyev chart. There are 92 non-radioactive elements. Do they all have a role? In about 400 years we will know, if the present rate of discovery is continued.

It is humbling indeed to realize that the animals already know.

We know about zinc and some 200 enzyme process the element accounts for.

We know about copper and healthy blood cells.

We know manganese is necessary for conception.

We know boron is associated with the parathyroid gland function.

And we know about a few others, but mostly the best settled science is still in the dark about what the cow has been trying to tell us.

Jerry Brunetti has a matchless summary for the topic of diversity:

The Soil Connection

It is my responsibility to alert the reader that this discussion does not address forage quality and pasturing success as it pertains to sound pasture management. This of course includes managed intensive rotational grazing, with adequate rest periods for recovery, etc. Nor does this discussion fully address soil fertility and agronomic practices necessary for optimum forage quality. There are soil fertility parameters that have a direct correlation to the nutrient density of forages, which in turn are necessary for livestock to be productive and healthy. On soils that tend to be imbalanced in poor fertility, or both species diversity — including deep-rooted herbs — can assist in bringing up fertility from below and hastening the decay process in order to recycle nutrient residues associated with urine, manure and forage, both foliage and roots. This can be helpful especially when the soils in question are natively deficient or depleted from abuse or neglect, and the economics of purchasing fertility from off-farm sources becomes a prohibitive option.

Starting with soil fertility, the model developed by William Albrecht, Ph.D., has a long history of success, utilized on hundreds of thousands of acres with a wide range of crops. Using a method that incorporated what is known as base (cation) saturation, the goal is to provide a saturation of the soil colloid comprising: calcium, 65-75 percent; magnesium, 12-15 percent; potassium, 3-5 percent; sodium, less than 3 percent; phosphate levels (P_2O_5) should be in the range of 250-500 pounds/acre, sulfur, 50-100 pounds/acre; boron, 4-5 pounds/acre; copper, 4-10 pounds/acre; zinc, 10-20 pounds/acre; manganese, 50-80 pounds/acre; and iron, 100-150 pounds/acre. These numbers of course are ranges dependent upon a Mehlich III Extraction Method and certainly allow for some flexibility.

This information is provided to note the relevance of forage quality and is hardly meant to be a synopsis on the concerns of productive soils. Most nutritionists used a wide range of lab determinants to gauge quality. My first inclination is to look at the mineral levels to see if I'm "on target," *i.e.*, certain mineral levels and mineral ratios give clues as to the quality of protein, the presence of energy, the ability of that forage to supplement an animal's needs for immunity and reproduction, and so forth. If the minerals are absent, I am suspicious as to whether this forage can supply the necessary essentials for productivity and health, regardless of the crude protein or relative feed values. Of course, the "proof of the pudding is in the eating," and ultimately livestock will prove the quality of their forage based upon production, reproduction, immunity to disease, healthy offspring, milk and meat quality, including flavor, keeping and cooking characteristics, and so forth. Keep in mind that typical soil and forage analyses often do not test for all the critical trace elements required by livestock, including selenium, chromium, cobalt, iodine, silica, vanadium, etc. This fact makes a strong case for diversity, especially of deep-rooted plants, which lessens the vulnerability inherent in forage that includes only a few species that, although efficient in accumulating certain minerals, would be inefficient in accumulating others.

The following table lists the levels of various minerals associated with a productive forage.

Mineral Content

For domesticated forages, having calcium levels approaching 2 percent provides a superior quality of protein than that of forages with less than 1.5 percent. Additionally, high calcium levels indicate forages rich in energy, synthesized as calcium pectate. Although crude protein levels are preferred in the 20-22 percent range (or 3.3-3.5 percent nitrogen), sulfur levels should be at least 10 percent of the nitrogen. That is because a 10:1 or lower nitrogen-to-sulfur ratio indicates that there is less non-protein nitrogen (NPN), and therefore the protein con-

Targets for Conventional Forage Quality

Nitrogen:	3.50 percent
Calcium:	1.60+ percent
Potassium:	2-3 percent
Magnesium:	0.50 percent
Phosphorous:	0.50 percent
Sulfur:	at least 10 percent of Nitrogen level
Chloride:	0.40 percent
Iron:	<200 ppm
Manganese:	35+ ppm
Copper:	15+ ppm
Boron:	40+ ppm
Zinc:	30+ ppm
Aluminum:	<200 ppm

tent has a more complete amino acid profile. Sulfur is also a vital component of the essential amino acid methionine, as well as cysteine, precursors to glutathione, a tripeptide antioxidant that also happens to be a building block of glutathione S-transferase, an important liver detoxifier, and glutathione peroxidase a critical immune activator. Phosphorous is a necessary element of ATP and ADP, energy molecules associated with the Krebs Cycle. Magnesium is associated with over 300 enzymatic reactions, including energy production in animals.

Trace element deficiencies, quite common in today's conventionally grown crops, are associated with soil depletion, soil erosion and hybridization. Volumes have been written on their multiple catalytic properties, so necessary for immunity, reproduction, growth and performance. Zinc, for example, is associated with at least 200 enzyme processes in the body; copper is a component of healthy red blood cells; manganese is absolutely necessary for conception; boron is associated with the parathyroid gland. These comments address just a few of the many elements necessary for optimum health and production, and we've barely begun to list their numerous functions and benefits as they relate to profitable livestock production.

Conclusions

Incorporating plant biodiversity on a livestock farm increases the diversity of animal-required nutrients, including minerals, vitamins, pigments, enzymes, amino acids, fatty acids, sugars and other carbohydrates, sterols, hormones and the numerous phytochemicals that are able to provide countless medicinal and metabolic properties. Increasing the farm's plant biodiversity provides weather-proofing from heat, drought, frost and excessive moisture. It minimizes the vulnerability that monocultures face through the vagaries of weather, because different plants have different strengths and weaknesses with regard to climatic influences. Complex plant polycultures also create numerous microclimates, which are able to buffer the extremes of temperature and moisture. Shade from trees and hedgerows can offset production losses associated with heat and humidity impacting live-weight gain and milk production. Windbreaks can reduce winter feed requirements by effectively reducing, even eliminating, the "wind-chill" quotient.

An extended food supply can be more readily realized with a biodiverse livestock operation, starting with early growing grasses, legumes and herbs, then later-arriving leaves, and finally berries, fruits and nuts late in the season. Woody plants have the advantage of actually having a year-round growing season, thus proving more efficient than grasses and certainly row crops in producing biomass. Winter browse on terminal buds provides exceptional medicinal components and a high level of nutrient density.

Plant diversity also increases the diversity and number of other wildlife, including songbirds and bats, which consume insect pests affecting plants and animals. These in turn attract raptors, which then prey upon rodents. Pollinators and predatory insects are able to find habitats and in turn help increase yields of crops bearing seeds, fruits and nuts. The soil food web, or soil ecosystem, is enhanced due to a permanent polyculture of plants growing on undisturbed soils. This means more efficient nutrient recycling and healthier root systems for all plants, again contributing to farm productivity. A healthy polyculture also means improved water percolation and purification, translating into cleaner groundwater and surface water, devoid of silt and excessive nutrients, and this situation ultimately benefits the ecosystems of invertebrates and fish in streams and lakes.

Plant diversity with livestock can readily provide the opportunity of two or three income streams for the farm, while also improving the farm's health. Animal products such as livestock, meat, eggs and dairy products; the use of timber as lumber or fence posts; fruits, nuts and berries to offset purchased feed and/or sold directly to the human marketplace — all offer multiple economic rewards that don't necessitate additional (net) human labor investments. This is especially true when factoring in the reduction or elimination of conventional agricultural practices and/or equipment.

15

The Down-Under Story: Unpaid Microbial Workers

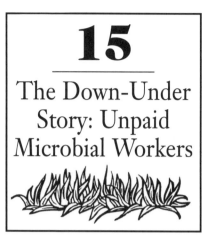

Grass, forbs and forage are nutrient uptake machines with an enviable record for efficiency. If the nutrients are available, that is inorganic elements are made organic by the action of microorganisms, grass is the most efficient plant for effecting uptake. Forbs and so-called weeds are specialists. They tend to concentrate nutrients. There are five functions only the life in the soil can perform. No liming program, no N, P and K application, no rock powder can do these things. Only soil life has a handle on soil life function, its reason for being in the food web. The term food chain is out of favor nowadays. The term of choice is food web.

The word "decay" invokes a biological process, not an organic chemical process. Gypsum will not decay corn stubble, for instance. Microorganisms are the decomposers. They have names and classes. There are fungi, bacteria, protozoa and a grand miscellany of life calculated to outweigh all life forms above the ground. When the decomposers are present and have the right air and moisture, they easily reincorporate crop residue back into the soil and pave the way for nutrient uptake into the root hairs of forage.

Subsoil drama is never one-dimensional. Nutrient retention in terms of nitrogen in the recipe. Once nitrogen has been retained, a pool is constructed as protein which is not available to the plant.

In order to get that nitrogen back into an available form at the rate of one pound per acre per day, figuratively speaking, this is a meal a day for the crop. All this translates into being a function of the food web called nutrient recycling.

In terms of plant protection against viral, bacterial, fungal and insect crop destroyers, the web constructs its balanced hormone and enzyme systems. Nature's own best example is the rain forest of Panama or South America or even the one on the Olympic Peninsula of Washington. If nature's own protective system can be classified that way, the real fungal activity is seated in the bio-terrain of the plant. Crop destroyers are invited only when the life system has been short circuited, first by malnutrition, then by rampant chemical intervention.

Great stands of grass and forage built over years seldom admit disease as a runaway thing. If destructive organisms exist in such grassland swards, they are kept well in check. Root rot fungi, root rot bacteria, parasitic nematodes all submit to biological control, seldom to chemical intervention.

When the microbial part of the food web is functioning, the foundation of control is in place, meaning root rot, drop-off, and other soil borne diseases.

Grass, with its mycorrhizal connections, even its lone DNA in a pasture, is seldom the victim row crops become. The answer is the food web in biological control.

Calcium, the prince of nutrients, flocculates deep. Polysaccharides, gums and gels generated by favored bacteria glue clays together. Fungal strands in turn wrap the subsoil package into being a water-soluble soil aggregate. The point here is that aggregated soil structure under a pasture cannot be achieved with soil chemistry alone.

Also emanating from the soil structure are certain kinds of plant signals. These signals are transmitted by plant hormones and enzymes. The message invites roots to a meal.

These signals are called exogenous. They are not produced by the plant, rather by the pharmacological capabilities of the soil microbes. The plant finds these signals and accepts the stimulation. The bonus sward of grass thus emerges as a product of microbial action.

The word is now extant that soil life easily dismantles pesticides. This is true to some extent if the preparation contains no chlorine, such as glyphosate. The food web organisms have the enzyme clout required to attack and breakdown toxic molecules. All in good time. If the food web is not functioning, then the clean up chore languishes. The business of recycling nutrients, restoring carbon is an essential part of pasture management.

The late William A. Albrecht once told reporters that the best fertilizer was a similar green crop turned in so decay could refurbish the soil food supply for the next crop. This is the best substitute for crop rotation. Grass and forage stand aside when rotation discussions take place. They build life under the surface while at the same time exhibiting their season's results for grazing and harvest. Jim Barlow and Elaine Ingham have discussed the food web connection at *Acres U.S.A.* conferences through the years. Those refined presentations apply to all crops even though abstracts in depth presented here emphasize grass and forage.

Elaine Ingham is co-author with a number of USDA scientists of the *Soil Biology Primer*, a USDA publication richly larded with illustrations of soil denizens such as bacteria, fungi, grazers, microbes, minerals, nematodes, protozoa, and trophic levels such as photosynthesizers. Organisms that consume photosynthesizers also make up the second level.

A healthy food web dismantles what some lawn afficionados call thatch. The inventory recited above decompose fallen blades, leaves that arrive from nearby trees, cow patties, even organic debris blown in on the wings of the wind. If the pasture does not police up the pasture, there is something wrong with the food web.

Any clutter on grass that inhibits aeration delays decay, not a problem in a well-managed pasture. Air, temperature, moisture all contribute to this recycling process, for which reasons cold and winter make the decay process take a bit longer.

Just the same, as soon as snow covers the surface of the soil, frozen acres come alive again. For this reason decomposition rates are often high under snow cover conditions.

Algae, yeasts, protozoa all do their work without a strike or labor battle. Ingham suggests leaving litter on a soil's surface. The hyphae connection to a nitrogen source establishes a decay mechanism soon enough, making the carbon-nitrogen connection. The minute carbon makes it into the soil, bacteria get a head start. Their enzymes are much more efficient at using this, the green stuff in plant litter materials. Notes Ingham, "Leave it as litter on the surface if you want fungi, plow or mix it into the soil if you want bacteria."

Bacteria is labile carbon material, the simple sugars, green litter — all are food for bacteria, that is why the mild dressing with molasses or sugar are bacteria food. Fungal food are woody type materials. Dry leaves qualify. The Florida water buffalo grower who permits live oak trees to rim his pastures invites trees to draw up rare nutrients and then favor the pasture with leaves for fall and winter decomposition. The C:N ratio for bacteria and fungi is 5. The ratio is up to 40 for fungi. The C:N ratio for green plant material is 30 to 100. Sawdust is 500 carbons for every nitrogen. Only fungi can hope to attack that type of material. This issue is fungal food or bacterial food. What does the pasture soil require?

Fungi vs. Bacteria

Carbon residue coming down from green plants becomes wonderful food for bacteria. Fungi also are a byproduct — if such a word can be used — of that food shower. The common denominator is decomposition.

Striving for a certain pH is quite beside the question. The soil life determines pH, not vice versa.

Bacteria tend to construct alkaline byproducts that slime. Therefore a dominance of bacteria will push the pH to an alkaline level. Take a parent material, pH 5, says Elaine Ingham, bacteria in the soil will push the pH to 5.5, even 6.0, finally 7. Bacteria can even push soil to pH 8.

If a fungal dominance becomes established, the pH will become more acidic. Fungi, of course, construct organic acids as their dominant metabolic product.

Soil organic matter is processed thousands of times by nature's unpaid workers. The various subsoil life forms consume each other much as the general work force consumes plant residue.

The scientific handle on the process can be illustrated with fluorescein diacetate. This is an activity stain, taken up by active organisms for transport across cell walls, and eventually, fluorescein accumulates within the cell. Under the microscope a count becomes possible, Withal, the investigator can find bacterial colonies on the surface of organic matter actively metabolizing measurable amounts of CO_2. Activity proceeds 24 hours a day under the proper conditions without a strike or faltering of zeal as choreographed by nature.

It takes 250 magnification to see this busy work that makes grass and forage production possible. Small is hardly adequate as a word to describe the scene. There are 600 million organisms per gram of the kinds of soil discussed here.

A 1,000 magnification makes each bacteria in a colony take on measurable length and width, enabling conversion of numbers to biomass.

Unless they avail themselves of an Elaine Ingham lecture or enter a laboratory, grass farmers are not likely to ever see single or clustered molecules of the soil microorganisms that make grass, meat and milk possible, first by transforming sunlight, next by soil organisms acting on inorganic minerals to make them organic.

This examination of bacteria and fungi is more than an academic exercise. The mere conception of fungal hyphae being measured — say, 16 micrometers, 2.5 micrometers in diameter — then and juxtaposing it next to a blade of grass makes the connection as mysterious as electrons orbiting an atom. Grass, much like a human being, is rented and occupied. Without once-resident micro-life forms, in the gut as in the cells, we couldn't drum a finger or write a line. Even the term *micrometer* boggles the imagination, take a metric stride, divide it into one million parts. One of these divisions is a micrometer.

This ability to measure explains why eyeball examination is always flawed. Fields and pastures will have inventory disparities from paddock to paddock, even within a paddock. Both sizes and populations will vary.

Pathogen, decomposer, or beneficial? The question haunts and infers. Heraclitus had its correct. Nothing remains the same. The mycorrhizal fungi on a root grows. Enlarged to the size of an Indy 500 car, it possibly travels and grows as fast.

Mycorrhizal fungi colonize all of one row crop, mustards excepted. Plants such as cauliflower, broccoli, the mustards, brassicas, all reject mycorrhizae. That is why corn can't follow broccoli in a rotation. The broccoli will have annihilated the required mycorrhizae.

The off-root drama beats anything man has ever put on the theatre stage. The hyphae go out from the root into the soil on a nutrient hunting trip. They carry the nutrient back to the plant in exchange for a carbon fix.

A mycorrhizal root, challenged with a pathogen, constructs an impenetrable defense. The protection mechanism wins, and the plant — both below and above ground — is saved from certain death. The mycorrhizae have to be there first. If arrival is tardy or concurrent with soil penetration, the pathogen wins. Why? The pathogen is virulent. The mycorrhizae are slow growing. The pathogen comes on like a first-magnitude star. It lights up the scene, does its damage, then falls exhausted. The mycorrhizal defense is that of a tank crawling slowly, sweeping aside hotshot organisms that have been programmed to assert themselves.

The art of innoculation was developed to give transplants a head start. With mycorrhizae instilled before a plant goes into the soil, pathogens find a checkmate when they arrive for a meal.

Improvement of the food web structure thus becomes the open sesame to every aspect of pasture management. This structure enriches the grass, presides over mineral uptake and finally makes grass quality preside over that traditional objective of having a sire settle 90 percent of the dams during the first 30 days of the breeding season.

Nematodes, like the making of books, have no end. Elaine Ingham describes nematodes with all the excitement of a Super Bowl game.

"The question is, can you see nematodes with the naked eye? Yes, there are pencil-tip sized nematodes one can see with the naked eye. Unfortunately, they are transparent. In dark soil they hide. Extracted, they can be counted, identified, in terms of genus and species. There are bacterial-feeding nematodes. The tribe has a pumping mechanism in their filter feeder. They find residence on the surface of soil particles on plants. They basically vacuum off bacteria. If a bacterial pathogen starts to grow on a root system, bacteria-feeding nematodes will vacuum them right off. What we see as slime is excellent food for them."

There are fungal-feeding nematodes. These little critters have a spear they use to puncture the cell wall of the hyphae. They suck out the contents.

There are root-feeding nematodes. There are ectoparasitic feeding nematodes that attack the root on the outside, and endoparasitic nematodes that penetrate the root cells in order to extract the internal contents. All root nematodes have a sort of jail hammer for the purpose of penetrating what to them must be as hard as we see concrete.

In her lectures, Elaine Ingham paints word pictures and she screens blowup photos that test the imagination of H.G. Wells. A fungal feeder is differentiated from a root feeder by the size of the stile. A root feeder looks like one of Hannibal's troopers taking apart one of Fabius' soldiers. Nematodes are specialists. Inside a cherry tree female, *Pratylenchus penetrans*, her young inside her, bursts open. Larvae chew their way out of the root, survive a couple of days in the soil, then return as endoparasites to attack the root system.

Predating nematodes are nematodes that eat other nematodes. These have buckle jaws as large as the jaws of a construction dragline, proportionate to their bodies. They simply put the target nematode into their mouths. The victim disappears like a string of spaghetti. If longer than the predator, the predator attaches itself to the side of its prey, then scrapes away until it breaks through to suck out the internal contents.

The soil is almost as scary as the back streets of New York. All sorts of characters eat and feed and set up the foundation for grass, that ubiquitous forgiveness of nature, to thrive.

Matchless drawings of all the above and more are a part of *Soil Biology Primer*, albeit without much attention to grass and forage. Just the same, organisms are mobile, as the pasture manager attempts to move the several tribes around.

Bacteria inserted in a soil stay put unless moved by water or a ride is hitched on some other life form. Arthropod hot spots can annihilate a section and leave another section undamaged. Some phyla prefer to feed on fungi. If the fungal community is strong enough, roots get a free ticket to survival. The fungal colonies become food web food for the arthropods.

When food web eating takes place, the resultant waste becomes excellent food for ammonium release as well as carbon. William Jackson's great book *Organic Soil Conditioning* defines "poop stuff" as a part of the carbon connection. Good tilth, in short, is the "poop" of organisms.

Tillage mixes carbon into a nitrogen rich soil, setting up a battle field in which the bacteria will win. Tillage cuts to ribbons the strands of fungi, decreasing the mass, turning it into a bacterial dominated soil.

The usual procedure for turning pasture into soybean ground is to till, to send CO_2 into the air, to wipe out fungi, to in effect mine the "poop" out of the soil. As a consequence, the organic matter level drops season after season. Residue returns organic matter to the soil, but it cannot measure up to the starting point. As Edward Faulkner pointed out in *Plowman's Folly*, the organic matter obeys the law of diminished returns. The plowing events since settlement of the Great Plains finally ushered in the dust bowl of the 1930s and now the lowering of the Ogallala Aquifer as wheat follows wheat, and circles of irrigated crops provide an aerial picture of the High Plains, little else.

There comes a time when the specter of desertification rears its head. The next step is either hydroponics or soil restoration.

If a century of grass production in the Texas hill country or on the High Plains is any indication, the pasture with herbivore traffic is self generating, self preserving, self mineralizing, as other sectors of this book readily reveal. Most soils are now in intensive care in the U.S., for which reason circa 1949, Poison Control Centers were established, and agriculture — pastures included — went to pesticides.

Most people think of earthworms as decomposers. Actually, they are not. Earthworms do not make the enzymes needed to breakdown plant residue. In fact, they live off bacteria and fungi in the soil and protozoa, nematodes and micro-arthropods.

The earthworm as a decomposer suggests an atomic reactor. Actually, the gut is a fermenter. The excretion is really an inoculant replete with bacteria and fungi, for which reason a burst of bacterial and fungal growth tracks the castings. It is therefore the function of the earthworm to help reconstruct the food web.

Pasture was, is, and always will be the king of no-till because it is a bacteria-dominated crop. The plowman's folly is more a folly when rototillers are used to pulverize the soil.

The stake of the pasture now invites a look at the developments new science has bestowed, starting with compost and compost tea, biologicals, microbial fixes, pasture feed and all the mineral nutrients pastures treasure.

In an agriculture that has elevated row crop production above the value of grass, consultant emphasis tends to favor bins and bushels, not the sward that maintains animals. Yet not one of the consultants interviewed for these pages dares suggest that pastures maintain themselves. The veterinarian Dan Skow and his associate Wendell Owens admit that some of their finest results are achieved when forage is given an assist from the eco-farming technology that has turned its back on simplistic salts and toxic rescue chemistry. Bill Lashmett and the Farm for Profit program exhibit tools and procedures that reward grasslands as well as row crop acres. Not one will suggest that providing trace minerals is of no consequences in the production of meat protein.

16

A New Zealand Connection

If there is a Mecca for grazing it is New Zealand. Next to John J. Ingalls' paean to grass, recited earlier in this volume, the living monument for grass and grazing culture is New Zealand itself. Alan Henning, a Fulbright scholar who journeyed down under for on-scene study of grass and grazing, came back to Wisconsin. He returned with a paean now his own.

"Grass is a natural health promoter," he said, adding, "it is also a stress reducer."

New Zealand is the grass country. The idea of plowing under the nation's green carpet in order to plant soybeans "for greater protein production," ADM asserts, is at best laughable to New Zealand graziers. There are fruit and boutique crop exceptions, of course, but in the main the two islands are wall to wall livestock. Henning reported back that there were 65 million sheep, 4 million dairy cows, 12 million beef cattle (and only 3 million people).

Grass really means forage, and forage means a mix of plants with a mix of nutrient and medicinal factors that still invites basic research if not DNA identification. Perennial rye grass and white clover often dominate the grassland swards, not as a checkerboard presence, but as far as the eye can see.

Grass management practices down under reject, if they do not entirely condemn, conventional farming practices in North America.

In talks to several audiences, I have often recited the story of crazy Ivan, usually to make the point that discussion and cooperation are the keys to success in farming, row crop, cow-calf, grazing, or specialties.

God came to Ivan. "Ivan," he said, "you are a favored son. Your work enables the sun to deliver solar revenue to plants. Chlorophyll and photosynthesis together with your labor and management feed the people. Therefore I have decided to grant one wish. If you want more acres, I will give you as many additional acres as you desire. If you want more money, give me your wish and the rules will be yours. If you want better production per land unit, the wish will be granted. If you want the grass in your pastures to be perfection, perfection will be yours and bloat and feeding distress will forever be banned. But know this, whatever you wish, the wish will be doubled for your neighbor, Vladimir."

Ivan thought this over. He would be happy with better pastures and crops. He would be even happier with more land. And more money. But to see that scoundrel Vladimir get double, Ivan couldn't live with that. Then a light went on in his brain. Suppose disaster came to Ivan. Then Vladimir would have double the disaster. A smile brought new life to his weather-worn face. He returned to face God.

"God," said Ivan, "put out one of my eyes."

If Alan Henning is read correctly, the American pastime of watching the fencerow to assess when a neighbor might go broke has eluded the New Zealand grazier.

"It is the power of discussion groups that makes New Zealand agriculture go," Henning told an Acres U.S.A. Conference. Some 35 such groups answer the questions and explain the answers.

The leading grazier experts are dairy farmers. Their job, the hardest in agriculture, is turning grass into milk. This act separates the dilettantes from the husbandry artists.

This habit of consulting with neighbors includes once-a-month comparisons of financial statements. The business of comparing vet costs, problems of every type, then soaking up answers makes competition real and strengthens the industry at the same time. These open forums for dairy farmers unveil opportunities.

Every two months meetings move to different districts. They often include farm walks, then problems that have surfaced during the time frame since the last meeting are solved — one at a time.

Bloat, breeding management, water supply — and the welfare of grass — all command attention. Clearly, a problem is an opportunity to close ranks and work together. The last scene of all during cooperative meetings is farmers making individual summaries of possible action. The next month the first to make a presentation is the man with last month's problem.

"We were all researchers," Alan Henning summarized "We were researching every day."

Breaking bread with fellow farmers, taking and giving prizes, overseas trips, all figure in maintaining *esprit de corps.* No doubt this civility within the ranks figures in New Zealand farmers being treated as first class citizens.

One lesson comes shining through from down under. Nature must be mimicked and obeyed at all times. Nature covers the earth with grass. It is up to the farmer to make economical use of this gift. New Zealand needs more land and American lands need more cattle.

To suggest that this simple key answers all grazing problems is to pursue wishful thinking. Just the same, Henning's New Zealand venture caused one principle to be imported back to Wisconsin. Walk the grazing area every ten days in every single paddock. This rule holds for every part of the United States, Canada, or any other country. This walk, says Henning, causes ideas to percolate and cascade. It anoints the thinking machine with information so fine tuned it bests the best computer.

All the top grass farmers in New Zealand do this.

The starting place for real grass farming is zero costs. The dictum Henning's mentor stated was that when you eliminate a cost, it will never increase again.

No cost, low costs, cost — this sequence reveals the degree of difficulty one encounters in getting rid of a cost. See what you look at is more than an aphorism. It is the key to rating costs, counting assets, and converting benefits to utilization and value. The pasture walk teaches one to compute the value of a cow pattie and the penalty for failing to achieve utilization as defined by Truman Fincher elsewhere in this book. Observation teaches the cowman something about species of green forage, grass, forb or legume, and takes the greater mix to be a greater asset. Discussion groups are probably more helpful than manuals or encyclopedias.

In Graham Greene's novel *The Heart of the Matter*, the character Scobie writes down the high temperature of the day in his diary. The idea is capital, as is a notation on the weather and every other observation while walking the pastures. Alan Henning counsels power walking the farm and making more than cursory observations. He picked up the habit in New Zealand while under the tutelage of a real professional grazier. Trees, fencing, cattle, the water supply, all demand constant scrutiny. The checklist of things to be done best takes form on the farm walk. The practice became a game down under and is now a stateside habit among those Henning counsels.

The feed assessment has to become first, not second nature, always in terms of dry matter, tons of silage, bales of hay, grain in the bin, whatever. Always, how many pounds of grass is out there?

It is amazing, Alan Henning mused before one Acres U.S.A. audience, how few farmers can actually compute the amount of forage they have alive and well on the farm. Sometimes the farmer has double the amount of feed he thought he had, the power walk making the revelation. Without a competent computation, pastures remain understocked and the magic of fresh grass is compromised.

Henning counsels caution when scoring grass. A division into thirds of available pasture calls for "ready to be grazed," "just been grazed," and "in between," all in terms of ten acres each as an example.

Call it staircase grazing! The point in question is to retain control of the pasture. When spring grass is difficult to control, why fertilize and exacerbate the control problem?

A rule of thumb is this: When you start utilizing about 80% of what you grow, start thinking about fertilizing, but don't put anything on yet. Putting the brakes on impulse is called "no cost."

A student of the international grazing scene, Henning is adamant when he says the principles are the same for the mountains of Monterey, Mexico and for the 80 acre plot in Indiana — thousands of acres or 80 acres! The grass is there to be utilized, not to make picture postcards.

The key to grass and livestock management is the rotation Henning counsels a longer rotation. "The upper Midwest calls for a 200 day rotation, roughly. If you have 200 acres and utilize an acre of grass a day, and the livestock on that acre need more feed, just plug in some supplementary feed. But maintain the rotation."

Rotations are not fixed in hardened concrete. They are there to be speeded up, slowed down, changed on account of weather. The full miscellany of farm considerations are distilled into the decision.

There is a grass budget, and there is a financial budget. There really are no absolutes including this one.

Grass is a cost quencher, one that many dairy operators have mislaid or forgotten. Dairy farmers seem to inhale costs, feed costs being over half of the budget.

There are answers. Put the dry cows on grass. Still the average dairy farmer in the U.S. spends $7,000 to $8,000 a year in bedding. This cost is calculated without adding the cost of laying it down, picking it up, spreading manure or storing. If you don't feed dry cows, the saving is absolute. Henning has an expression. "For grazing management must open the gate in your mind and then open the gate for your livestock." The basic problem with most graziers is preconceived ideas. They float around freestyle. They do not flow from facts and observation.

When power walks and observation order up change and new education, you, the family, the farm have to do all the above yourself. One part of the team cannot outdistance the other.

Cows will graze through snow, but they have to be trained. When they get their ration of good forage they stay healthy and maintain condition.

It may be that the American bison is the king of graziers. The animal survived blizzards and adverse conditions by grazing in large groups, breaking through snow and ice to graze. Theirs were a long rotation. Grazing through snow is about 50 to 60 percent efficient.

Uncommon good sense tells the farmer to narrow the raceways so cattle do not expend more energy than "straight ahead" demands. Shrubs and windbreaks do not need to be prefab capital guzzlers. The shelter often is not necessary. Calves and cows and bulls know enough to present their backs to a storm, to hunker down and construct a zone of comfort whatever the conditions. Usually the storm passes and the cattle keep on grazing. A little stress seems to be more important than perceived comfort.

New Zealand has provided the world with the barrel feeder. Hundreds of calves can be raised with this device.

A Henning dictum is, "*concentrate on the weakest area.*" The consequence will be a dramatic increase in average forage production. This is opposite the conventional ignorance which tells farmers to concentrate on their best area. Once the farmer concentrates on the worst areas, they tend to become the best areas and the previous best areas are enhanced even more.

Withal, different aspects of the property ask for special utilization. Areas that need more organic matter suggest themselves as feeding areas. South facing slopes suggest winter. North facing slopes, kiss them goodbye until spring.

What, then, is the bottom line when grass is the chief asset? Alan Henning has a few grazing cows and chickens, and he markets his own cheese label called "Jersey Grazing." He does Havarti, Gouda, Muenster. He sells to restaurants and farmers' markets. There are cheese curds and brown eggs, there is ground beef, there are roasts and smoked chickens. Cow-manure compost is also a product. Never considered is growing grain for Archer Daniels Midland. Grass is king, the way cotton was once proclaimed king. If the journey starts with a single step, then selling cheese starts with selling the first pound, according to Henning.

Finally, chickens on grass are mortgage reducers and promoters of grass.

17

Health, Pasture & Feedlot

Ted Slanker expressed a clear understanding of the mandate for pasture fed and finished beef, earlier in this book. Therefore the points entertained in this chapter stand as a mere codicil, an enlargement of the subject that cries out for attention in the wake of a bovine spongiform encephalopathy outbreak in the United Kingdom and Europe. Reports that British White, Normandy breeds, proved immune to the bovine spongiform encephalopathy (BSE), or Mad Cow disease probably had more to do with their pasture maintenance than other factors imposed on the debate.

There is a penchant in pseudo-scientific circles to sequester a single factor in a closed air-tight compartment for the purpose of analysis with little or no reference to inter-relationships of a thousand variables. The variable least considered seems to be the quality of the forage and the state of the pasture. On grass and forage that contains suitable fertility loads, the bovine's instincts will see to a diet that triggers hormone and enzyme systems capable of warding off viral, bacterial, fungal and even insect attacks.

Cowmen who have said "so long" to veterinary bills know this. Those who practice linear measurement selection and culling norms know it as clearly stated in *Reproduction and Animal Health*.

Hard on the heels of World War II, the United States Department of Agriculture went into an eradication mode for Bangs disease, or brucellosis. The scheme for eradication of brucellosis called for depopulation of reactor herds. A series of drought years and a marginal understanding of the nutritional requirements created a syndrome called infectious brucellosis abortus, abortion of calves in reactor herds being a consequence. Bangs was never a problem on fertility pastures. But, as happens with almost all cattle problems, a bug was identified. The Dr. Experts — to use a Fletcher Sims term — decreed elimination of "infected" animals in order to eradicate the disease, the term "infectious" having inoculated Dr. Experts and cowmen alike with quivering terror.

Albrecht scoffed at this judgment. "Infectious brucellosis abortus," he told me, "is about as infectious as the stomach ache."

Albrecht had been associated with a Springfield, Missouri physician named Ira Allison. And Allison's attention had been attracted by the cases of undulant fever, the human variant of brucellosis. The Albrecht-Allison findings were generally reported in those days as the miracle of the Ozarks. Materials are archived at the University of Missouri, where Albrecht was head of the Soils Department. The brucellosis story finally became a matter of record in a paper by J.F. Wischhusen, an abstract of which was published in *Acres U.S.A.* in November 1974.

Poverty pastures were the issue at the time because the "poor for field crops, good for pasture" myth was still afloat. As a consequence, malnutrition invited brucellosis into herds.

Either copper was missing or its uptake was prevented — an imbalance that would one day figure in the arrival of bovine spongiform encephalopathy, or Mad Cow disease.

The control of Bangs disease has degenerated to the rock bottom called "eradication." This is somewhat akin to shooting the victims in order to get rid of a nasty problem. Yet there was a time when control of brucellosis with prophylactic and therapeutic treatments of designed doses of manganese, copper, cobalt, zinc and magnesium, plus iodine separately, made history — both in respect to human beings and animals. J.F. Wischhusen told the story in a rarely read paper entitled "Brucellosis and Mastitis."

A few of those paragraphs merit abstraction in depth, lest we forget lessons learned before. Also, there is the reality that some farmers simply feed brucellosis out of their animals before the eradication sharpshooters get to them.

After two years of experiments, Wischhusen said both Bangs disease and mastitis were on the way out. Increased milk production, increased butterfat content of milk, better calf crops — all follow elimination of brucellosis and mastitis. "The cost of feeding a daily requirement of trace elements designed to correct brucellosis has come back to some dairy farmers 17-fold," Wischhusen noted.

But how? Here is the story.

History

Dr. Ira Allison, Springfield, Missouri, held his fourth clinic for undulant fever patients March 26, 1949. Of the 47 patients who registered and whose blood culture tests were negative, only about half could be interviewed, such was the interest from visiting physicians, soil scientists, animal nutritionists and other interested observers. The story of all patients at this clinic was the same as from those who testified at previous clinics. Where previous treatment had failed, Dr. Allison's therapy provided rapid restoration to good health. Not only was undulant fever controlled, but apparently the therapy was effective in cases of gastric ulcers and other stomach troubles. It is important to note that symptoms of nervousness, despondency, fear of imminent danger all disappeared after three months of treatment.

One patient, Mrs. E. Roby, age 79, had her troubles diagnosed as "incurable eczema," that manifested itself as scales on her hands from which blood exuded from ordinary pressure, for instance, when milking cows. She experienced complete recovery in eight months from Dr. Allison's therapy, and was naturally very pleased to so testify. Among other unusual cases present was a diabetic male, age 52, who after three months' treatment, was enabled to get along with one-third the insulin formerly required. A boy, age 12, suffering from tularemia, became well after four months, and had his teeth improved. The absence of trace elements, it can be seen, leads to many disorders and diseases.

Soils on which food crops are grown have been depleted through a century of cropping. Trace elements have never been commercially replenished to our soils. Specific applications of manganese, copper, zinc, etc. are only made in a few territories where acute visible symptoms have made this mandatory; cobalt and iodine have never been applied to soils commercially anywhere. It is now known that plants often do not contain these necessary trace elements in quantities sufficient to provide amounts necessary to animal health.

Trace elements are a daily necessity in the diet of man and animals, as are vitamins, but it is apparent that trace elements, the biocatalysts, precede vitamins, and may actually cause their formation. This is now best demonstrated with cobalt, through the discovery that vitamin B-12 is a cobalt complex containing 4 percent of elemental cobalt. It goes without saying that vitamin B-12 cannot be synthesized in the animal body without cobalt, but also that an application of cobalt to soils and to feed will take care of the vitamin B-12 factor. In fact, when vitamin B-12 is destroyed by heat through pasteurization or in canning operations, it is the elemental cobalt that remains. Manganese too has been shown to be related to the synthesis of vitamins, particularly A, C and E. For instance, the effectiveness in breeding troubles of wheat germ oil that contains Vitamin E is related to the manganese contents of the germ of wheat.

Farms Under Bangs Experiment

On the 265-acre farm of the Nicholson family at Ash Grove, Missouri, on which a herd of 80 Jerseys was totally infected with brucellosis, the owners faced ruin if their herd had to be liquidated. They agreed to an experiment that involved fertilizer applications and feed supplements. The farm had been in the possession of the Nicholson family for 104 years. The experiment was made under Professor A.W. Klemme, Soil Specialist, University of Missouri. In 1947 they only raised 11 calves from 55 cows; in 1948 they had a calf crop of 38 from the same 55 cows. Since feeding the trace element supplement from June 1948 to April 1949, only 3 calves were lost, and those from non-infected cows. In the herd of 77 Jersey cattle, there remained three Bangs positive, with about five suspects. But Thornton Nicholson, one of the sons, said he had been getting bigger and better calves. Also cows were formerly troubled with soreness between the toes, but since feeding the trace element therapy, this problem cleared up. They definitely improved in health, George Nicholson, Sr., reported.

Final conclusions could not be reached on this experiment until soil treatments had a chance to show results via the crops, and blood culture tests were made of all animals. But in a preliminary way the work under controlled conditions confirms the practical results previously reported from other places. Brucellosis disappears when manganese, copper, cobalt, zinc and magnesium are either fed in a therapeutic dose or applied to the soil as a prophylaxis. The Nicholson farm is located on a typical prairie type of limestone soil in southwest Missouri, high in iron, not unlike the soil at Hagerstown, Maryland. It is rich in lime and iron, but poor in phosphorus. Animals in consequence require their needs of phosphorus to be brought into balance with calcium, before the trace elements, particularly manganese, can be expected to function properly as catalysts in bone formation, milk production and a strong and healthy calf crop. Phosphorus here would appear to be the first limiting factor, so that the supply per se requires attention. There were still some eight cows not settling as they should, and these had all the gross visible symptoms of a severe phosphorus deficiency. But early in April these eight cows too settled.

Soil-Animal-Nutrients Relationship

The matter of furnishing inorganic essential nutrients adequate for animal requirements through fertilizer applications to soils must be viewed within its well-known limitation. Plants readily assimilate potash and nitrogen from soil; but these elements are not needed by animals to a like extent. On the other hand, animals require calcium and phosphorus in concentration not adequately assimilated by plants. Furthermore, it must be remembered that animals (including, of course, always man), need salt in amounts, if it were expected to be furnished via soils, would destroy crops. Therefore, it can readily be seen that irrespective of good fertilizer practices, animal rations require supplements of calcium, phosphorus and sodium chloride. Furthermore, the potash and nitrates present in crops used for feed require adjustments to offset any toxic or inhibitory effects which by their excessive presence may make mandatory. Excessive potash may immobilize such animal nutrients as magnesium, sodium, calcium, so that to offset this high potash an extra addition of at least these three becomes necessary. We have yet to learn of a suitable antidote for the toxic effect to livestock of high nitrates in feeds. Animal nutritionists would probably frown upon the addition to feeds of as little as 5 parts per million of boron, yet alfalfa and other crops assimilate as much as 40 parts per million of this element, the effects from which are not known.

Criteria for Trace Elements

The criteria for the use of manganese and other trace elements by animals is now firmly established by the dosage required for combating brucellosis, mastitis and breeding troubles, and they amount to:

One ounce per 1,000 lbs. body weight per day of:

200 parts manganese sulphate 65 percent feed grade

20 parts copper sulphate powdered

3 parts cobalt sulphate monohydrated

1 part zinc sulphate monohydrated, plus suitable doses of iodine and magnesium according to local conditions.

The above concentrations are well within toxic limits; in fact, toxic amounts are so high that they have not been ascertained. Experiments with manganese in blood injections on rabbits showed that when related to cattle, the average cow would have to consume two pounds manganese sulphate at one time to be fatally toxic.

The good results obtained on dairy herds from the feeding of trace element supplements to feeds on a number of Missouri farms visited, were even better than the results so far obtained at the Nicholson farm.

Farmers who themselves personally suffered from undulant fever and recovered therefrom through Dr. Allison's trace element therapy were naturally anxious to feed their herds back to health by the same method.

Ovarian cysts noticeable in some cows were also cleared up, and this is explained perhaps by the fact that a manganese deficiency in heifers showed up in their ovaries. Reference is made to Wisconsin bulletin 474, January 1948, showing ovaries from a group of heifers on low manganese ration contained 0.85 parts per million of manganese, while ovaries from the manganese supplemented lot averaged 2.65 parts per million — a threefold increase. This bulletin also refers to the manganese concentrations in blood of animals which normally fall within a range of 4-8 micrograms in 10 cubic centimeters. In contrast thereto, semen from young bulls was found to have a manganese concentration of 26 to 80 micrograms in 10 cubic centimeters. This shows that manganese is concerned with reproductive functions, and it is no wonder therefore that a variety of breeding troubles are cleared up from its optimum presence. The function of manganese here is of course entirely apart from its action as a catalyst in food assimilation.

The multiple benefits obtained by Missouri farmers from the use of therapeutic doses of multiple trace elements may be summed up as follows:

• Better appetite after a few weeks.

• No mastitis.

• Brucellosis disappears after a few months.

• More milk (the average loss in milk production by Bangs reactors is estimated to be around 2,000 lbs. per annum).

• Lower bacteria count in milk.

• More butterfat; increases from 4.4 percent to 5.4 percent were general.

• No cystic conditions.

• Bigger and stronger calves (afterbirth promptly removed without artificial help).

• Much improvement in settling.

Other gratifying results on five different farms were obtained when feeding the mineral therapy to horses to correct fistulas as the result of Bangs disease. These cleared up and go to show the need of the same therapy for horses to prevent not only fistulas, but breeding troubles, the same as in cattle.

All warm blooded animals are liable to be affected by brucellosis, so that this disease is a criterion for their need of manganese, copper, cobalt, iodine, zinc and magnesium.

Brucellosis, mastitis and breeding troubles are generally treated with a prophylaxis or a therapy of manganese and other trace elements in doses designed for such purposes — happy discovery indeed. It bids fair to rid the country of Bang's disease and mastitis as effectively as perosis in poultry and other direct deficiency symptoms now under control.

It may be reasoned that wherever brucellosis or mastitis is present in any class of livestock, their feed supply is, or has been, out of balance as far as manganese, copper, cobalt are concerned. When this balance is restored through suitable soil treatments or feed supplements, the incidences of brucellosis and mastitis disappear. The work of Levis and Emery was conducted in Cleveland on cattle from northern Ohio, but Dr. F.M. Pottenger corrected undulant fever cases in California from a treatment based on their formula. Dr. Ira Allison, Springfield, Missouri, cured well over a thousand cases of undulant fever during the two-year period 1947-48, again substantially with the same therapy. This shows that over a wide area the same deficiencies are involved in brucellosis, and that a correction is possible with the same treatment in any locality. Here then are definite indirect criteria by which to judge the need of manganese, copper, cobalt, iodine, magnesium and zinc not only as plant nutrients to be furnished via soil or foliage, but as inorganic nutrients that can be furnished direct to animals and humans as feed and food supplements.

According to the physician, George H. Earp-Thomas, working at the beginning of the last century, the nation's soils as represented by New Jersey, had ran out of cobalt by the year 1900. Cows may get by without this nutrient by finding one of these substitutes we do not completely understand. Perhaps a trace arrives via the air. In any case, its absence sets the stage for brucellosis when accompanied by an absence of or a marked imbalance with magnesium, copper and zinc, as illustrated above. The solution Allison found is as simple as a fertile pasture, and has only to be tried.

"Not good enough for a row crop, but good enough for pasture" is still a worm that rattles around in heads used only for billboard-advertising absconders of agriculture. Quite the opposite is true.

Acute diseases are created, first by nutrition in deficit, then by genetics that often as not result from failure of forage in the paddock. Diseases are exasperated by confinement. Cattle on grass find the immunity nature has decreed, an immunity that falters and fails in the feedlot. The term "feedlot bloat" has no counterpart called pasture bloat when good grazing equals contented cows. The feedlot answers with ample rations of Arm and Hammer bicarbonate of soda, and that failing, the invasive lose.

Para-T or Johne's disease is alien to the dairy animal maintained on pasture. This syndrome is estimated to affect perhaps 60 percent or even more of confinement dairy cattle. The architects of control by annihilation of the herd have repeatedly asked Congress for over $2 billion for the purpose of Johne's eradication.

Eradication seems to be the banner of the opportunist, the boast of the simple-minded, or the mechanism of the politically dishonest.

Foot & Mouth

Hoof and mouth disease, now called foot and mouth by the USDA to comply with international word preference, has not appeared in the United States in any significant way since frontier days. At that time starved cattle came north out of Mexico carrying the aftosa virus. Old timers will tell you the infected cattle were too far gone to be saved even if someone knew the cure. There was a slight blip on the disease screen some years later, in 1938, and the shouting got underway.

Eradication is necessary, the Dr. Experts have it, because the disease organisms are so virulent they can't be controlled, not with medicine, certainly not by dealing with forage or the mineral box.

British White and Normandy cattle seem to escape the epizootics of the hour because they are *pasture cattle*, and their owners seem to think first of the welfare of their animals on quality pastures.

Grass & Forb Therapy

Grass and forage is the best medicine. It fails when industry showers planet Earth with levels of nutrients that become toxic when grass becomes overloaded — as is the case of manganese in cases of bovine spongiform encephalopathy, brucellosis and other syndromes little understood or forgotten accidently on purpose. The aftosa virus is a good for instance.

In the United Kingdom, the veterinary profession is deliriously happy. The European Commissioner has noted that they have identified 13 different strains of the virus, once categorized as three strains of aftosa viruses with varying virulence. The deadly cardiac type had the reputation for wiping out a herd in short order, some of the infected animals dropping dead of heart failure before they could develop lesions in the mouth or on the hoofs.

In his seminal work, *The Survival Factor in Neoplastic and Viral Diseases*, Dr. William Frederick Koch detailed such an epizootic, an abstract of which follows. The nature and cure of the infection was made a matter of record by the Agriculture College of Brazil.

There were 59 head of cattle and 200 pigs. The professor expected his herd to be wiped out, since five animals had died quickly. Other animals went down, unable to rise. Still others exhibited lesions in the mouth and feet, but were able to walk.

All of the live animals were treated. Fifteen were calves, 17 were young bulls, 15 were cows. All were considered favorable hosts for the destructive virus. Two cows and one calf died even after treatment. The rest of the herd were saved and exhibited no ill effects.

Treated swine fared as well, only 33 being symptom-free. Four animals died of infection, giving a cure rate of 98 percent. *Prevention for the uninfected was 100 percent.*

Immunity held for the three years the case was followed by Koch.

In another case, Dr. Alberto de Silva found Type C infection in the rural university herd, circa 1957. Here the viruses had proved active every year, usually killing 10 percent of the cows. Milk production declined as a result of the disease. All the animals were treated, with no resultant deaths. Within hours after treatment, dried-up mammary glands secreted milk at the normal rate. Cure factor was 100 percent.

At the Rubino Institute in Montevideo, Uruguay, a bull-calf experiment was set up. All were infected with a cardiotropic virus that killed 80 percent in four to nine days. The other 20 percent became chronic heart cases that died later on. The virus was 100 percent fatal in the dose used.

The experts prepared a vaccine from this virus and protected 10 cows with it. Those 10 were sequestered from another group of 20 cows, half of which received the Koch treatment, the rest held as untreated controls.

The bottom line: All animals but one recovered when Koch treated, even though infected. Of the vaccinated animals, four of the 10 died. The untreated cows also became terminally ill, and lacking the facilities to incinerate the doomed animals, they were given the Koch treatment. All survived.

The treatment was simple, yet complicated.

Viral Disease

The only way a virus can live is to find food. That food has to be toxic. It is the function of molds, fungi and bacteria to confer a suitable level of toxicity on human beings and animals. The molds run from white (mildly toxic) to red, which are lethal. A viral infection is not possible unless there is a suitably toxic host. In a manner of speaking, viruses are the key to the immune system. They have to utilize toxins, keeping in mind excesses are toxins.

In addition to nature's toxins, including those manufactured by the body itself, there are the manmade toxins, thousands of them spread across the Earth with reckless abandon over the past 50 years.

Whenever animals become too concentrated per square mile, or too many within a given area for any length of time, they almost always develop diseases because of toxicity buildup. Fecal and urine contamination by animals is about the same as an equal number of human beings without sanitary facilities, hogs excepted because they produce about twice the pollution, certain animals excepted because of size and feed capacity.

The supply of toxicity under the imprimatur of high science has exacerbated the natural problem exponentially.

Further, when you go beyond chemistry, beyond molds and bacteria, and get to the energy level, the mismatch in itself becomes a form of toxicity.

If a virus is present among the cattle, it is also present in the wild. Annihilating cattle en masse in order to create a disease-free area — carcasses remaining unburied or unburned for days or weeks — is merely an insurance policy for continued epizootic diseases. That's the word for widespread animal diseases.

We can accept the axiom that if there's food for the virus, the virus will generate *in the absence of oxygen.* The oxygen thesis was confirmed at the time of the calfloo and pigloo experiments shortly after World War II. With only an air curtain on the south side, as opposed to a doghouse opening, it was discovered that in a full complement of oxygen, a pneumonia virus could travel only a few inches to find a new host. With even 1 percent deprivation of oxygen, the virus could reach the far wall of a fair-sized room. A well-situated host and an operative transport system has been and remains the author of any epizootic.

Koch

It was this inventory of information that brought on a clash between the Koch remedies and allopathic drugs and the American Medical Association circa World War II. William F. Koch's approach utilized the oxygen connection. His formula relied on a homeopathic dilution, only one part of the remedial substance to one million parts of distilled water.

The medicine, the AMA asserted, was useless if not harmful. In *Survival Factor*, Koch explained that he was looking for something that worked. He was a physician, not a veterinarian, and therefore most of his experiences merely helped terminal cancer patients, victims of polio, and degenerative metabolic disease victims in general.

Koch's approach can be explained as follows.

A dilution of the effective oxygen agent of one part per million carries thousands of billions of molecules in each cubic centimeter, and only one molecule is needed to start a catalytic action which can grow geometrically — the "awesome chain reaction."

This hypothesis relied on the work of Michael Faraday, the father of electricity, who revealed that each chemical had an electrical charge. Koch evaluated the electrical properties of his chemicals. In describing the dilution required to produce his catalyst, he said the electrons are more active with proper dilution. If only one molecule was required to start a chain reaction, a molecule with electrons properly activated kicked open the door to effective oxidation. Increasing oxidation was the objective of the catalyst.

It was oxygen that could sledgehammer a disease agent into oblivion. The carbonyl group, with its detoxifying action, proved to be the best common denominator. Koch explained all this to the FDA. This brought the AMA, the pharmaceutical industry and the FDA up fighting from their chairs.

Koch, much like Harry Hoxsey, was proscribed. He was brought to trial not once but twice. A parade of over 200 people — including William H. Dow and William J. Hale of Dow Chemical Company — testified on his behalf. It mattered not.

Dr. Morris Fishbein, of the AMA, had editorialized in the April 18, 1942, issue of the *Journal of the American Medical Association*, claiming credit for having mobilized the FDA for the purpose of closing down the physician.

The FDA and AMA further mobilized the Federal Trade Commission. This regulatory agency challenged Koch's theory of toxin destruction by increased tissue oxidation. The agency asked the court for a federal injunction. The court did what most courts do. It ruled in favor of the government that pays its salary. Koch fled to Brazil, where he conducted the hoof-and-mouth programs.

Koch's treatment had a name, *glyoxylide.* Koch bubbled ozone through glycerine, then diluted the product to the effective level. Canadians used it effectively in acetonemia, mastitis, brucellosis, and animal disease in general, shipping fever included. Shipping fever, of course, has been eradicated by the bureau process of calling it IBR — infectious bovine rhinitis.

What is Going On?

While America cowers and Brits continue their nihilistic program much like a medieval pogrom, or an Eichman variation thereof, a few able and dedicated scientists continue their work, standing on the shoulders of giants.

The use of hydrogen peroxide in animal husbandry is one of the most ignored developments in the U.S. farm papers. Equally unreported is the political nature of the U.K. animal program. As one Brit put it, "If you attended the right school, have the right connection, you're excused from the law, ethics, even from having common sense."

Unfortunately, the Koch remedies are gone, extinguished by the handlers of the agencies set up to protect people and farms, but the lessons remain.

Oxygen is the sworn enemy of viruses. Technically, it is the O_1 nascent oxygen atom which is the oxidative fraction. That is the active medicinal element in hydrogen peroxide and ozone. Hydrogen peroxide is H_2O_2. Ozone is O_3. The particular benefit of ozone is the extra oxygen atom. O_3 splits into O_2 plus O_1, thus O_3-> O_1 + O_2, the extra benefit of ozone. Thus the protocol presented above. Interested readers are directed to Robert Stroud's *Diseases of Birds* for a look at the singlet oxygen.

Conclusion

It is impossible to suggest that officials are unaware of this science or its protocols and conclusions. The U.S. officials know the

Protocol for Treatment of Animals with Hydrogen Peroxide Against Viral & Other Diseases

by Alwyne Pilsworth

The prime requirement is to supply the cattle with drinking water containing 30 to 50 parts per million of 35 percent *food-grade hydrogen peroxide.* To achieve this, you would add 500 mls (17 oz.) of 35 percent food-grade H_2O_2 every 5,000 liters (1,330 gal.) of water. If disease attack is imminent, increase the H_2O_2 to as high as 7 liters (7.5 qt.) per 5,000 liters (1,330 gal.) water. There should be no adverse side effects whatsoever on the animals.

It is important to start with *clean drinking troughs,* otherwise the active component is quickly lost. The active component is the O_1, the nascent singlet oxygen atom which is attached to a molecule of water. It is strongly oxidative in action and readily reacts with and destroys viruses and animal pathogens.

The quality of the water is important for maintaining the content of active hydrogen peroxide.

A high-cation content of water with minerals such as calcium, sodium, magnesium, potassium, iron, etc., will react very vigorously with the singled O_1 oxygen atom, releasing oxygen slowly and thus remove the oxidative O_1 fraction, which is the active agent. For example, we found when spraying H_2O_2 on crops of potatoes that if the sprayer had been spraying fungicides, the active component of the H_2O_2 was lost within a minute. It is attention to details like this which is the secret of success with peroxygens.

It is also important to understand that when there is a high level of contamination of both cations and organic matter (such as organic acids) in the water, then reaction will also occur. Thus it is advisable to check the components of your local water supply before commencing any treatment. It may prove necessary to top up the H_2O_2 component at regular intervals to make an allowance for O_1 loss. It is more efficient to install a header tank fitted with a metering pump to maintain the optimum content of H_2O_2 and thus the O_1 oxidative atom. A little variability in the solution strand is relatively unimportant. It will not harm the animals.

It should be understood that H_2O_2 is perfectly safe for ingestion by all animals and that the same strength of solution is suitable for all animals.

Animals which have already developed the disease may be treated by infusion of H_2O_2 into the jugular vein at the rate of 1 cc of 35 percent H_2O_2 per 50 kgs (100 lbs) bodyweight. The H_2O_2 should be diluted in saline solution and introduced into the vein slowly using a large syringe and a small, 18-gauge needle. Experience has shown that infusion treatment is to be preferred for all ruminant animals.

H_2O_2 may also be given in feed. On dry feed, apply 5 liters (5.32 qts.) of 35 percent food-grade H_2O_2 diluted in 10 to 15 liters (2.66 to 4 gal.) of water to each ton of feed. For hay, the same mixture should be used, but add a little honey to the solution. Dissolve the honey in warm water to ensure a good flow.

country is rife with hairy warts on cattle feet, with downer cattle, sudden-death syndrome, ParaT — all spawned by toxic brews, aerosols and metabolic response thereto. How the political agenda plays itself out remains to be seen. For now, it is enough to note that five-way vaccines and seven-way vaccines are failing American farmers.

BSE Panic

Late in the 19th century, two investigators in Germany named Creutzfeldt and Jakob discovered a syndrome now named after them. It seemed to occur once for each million in the population. It still does, and usually is disposed of when the patient is buried.

In 1949, the United States set up Poison Control Centers. By then the decision had been made to take agriculture along the industrial route. Soil was just dirt, and poisons would be rated on an LD_{50} basis, meaning that such-and-such a dosage per unit of body weight would kill 50 percent of test animals. This simple-minded viewpoint scrambled all sorts of nutritional relationships as prions annihilated enzyme-producing microorganisms in the soil.

Writing some 50 years ago in *Soil, Grass & Cancer*, André Voisin noted how certain chemicals such as magnesium interfered with the uptake of copper even when that latter nutrient was available in the pastures. He called this "disease" *epizootic ataxia*. It sounds like BSE to me, and the university-blessed grinding of cats, dogs, cows, chickens and horses to make cattle-cake was not even a blip on the radar screen at that time.

First-rate panic has to have a totalitarian rationale. After Great Britain experienced BSE cases in animals, now identified by prions bent out of shape, the defenders of toxic genetic chemicals postulated an infectious disease agent of some sort and resultant transmissibility. The answer was to shoot herds and to construct a theory with a politically correct etiology. The only rule seemed to be: *Do not implicate the phosphate or manganese factories spewing their pollution into the ambient air, and never, never indict Phosmet poured on animal spines.* Hang in there with that tired bonemeal explanation, which brings BSE to cattle, but not to chickens or hogs, and probably camels are immune.

Suspicion hatched in the minds of bureaucrats became high science. Some judge rubber-stamped the government's case because judges almost always rubber-stamp the government's whims.

Mad Cow Disease

Slaughter of animals perceived to be guilty of incubating infectious diseases has an aged tradition. In *Evolution of the Veterinary Art*, J.F. Smithcors gives chapter and verse of European killings of animals conditions no longer a part of veterinary nomenclature. Most if not all of the conditions seem to have had their genesis in malnutrition. That cause combined with industrial pollution possible explains the horror story of the hour, transmissible bovine spongiform encephalopathy. The conventional explanation still a part of official rhetoric is that cattle cake, bone meal, protein bypass and feeds from sheep infected with scrapie transferred the "infection" to cattle, after which the agent was off and running, infecting human beings with an indestructable agent called a prion. That such an etiology has never been proved and that infection without an organism is junk science did not prevent it from becoming holy writ in the 1980s, 1990s and well into the 21st century.

In writing his eological detective story, Mark Purdey, an English farmer and scientist, tracked down the clusters of Mad Cow, chronic wasting disease and Creutzfeldt-Jakob disease — the last a human variant — in order to confirm and enlarge a theory he entertained from day one.

The first suggestion that presented itself was associated with organophosphates. Sheep were dipped in the stuff. Cows have the organophosphate mixture called Phosmet poured down their spines to battle grub. It seemed that a chemical pollutant made its way into the blood system, finally arrived at the brain and folded healthy prion turning the brain into a veritable sponge.

Then Purdey started traveling. He found clusters of the syndrome in Colorado, at the desert missile ranges, in Guam, Iceland, England, much of Europe, even in far flung Pacific islands.

Did well-managed pastures confer immunity to this syndrome? The question wasn't answered because it wasn't asked.

The first accepted premise that tumbled into a bottomless canyon of doubt was that BSE was transmissible by touch, even by proximity. Yet Purdey found a fenced herd of deer in New Mexico that was hundreds of miles from the nearest cluster. "How did the infectious agent jump the 500-mile gap between the longstanding hot spot in Colorado?" asked Mark Purdey.

Purdey found environmental factors shared by every cluster of the spongiform disease around the world.

The mining, distribution and presence of manganese oxide and wulfenite (lead molybdate) ores, the last containing copper-chelated molybdenum, declared their presence in one form or another wherever clusters of disease were found. Manganese contained in salvage from Japanese war planes from World War II may even have figured in New Guinea, where Fore tribesman suffered Kuru, a variant of Creutzfeldt-Jakob disease.

Purdey must have let out a long, low whistle as the implications settled in — as described in *Deadly Feasts* by Richard Rhodes. The successful transplant of brain material into a Rhesus monkey by D.

Carleton Gajdusek earned him a Nobel prize and settled for all time the transmissibility of the disease.

What was really transmitted? The manganese connection and the interplay of light and sound and magnetism must have made the difficult transmission possible, for which reason the fiction is afloat that not even 500 degrees Fahrenheit can kill the agent. In fact, it takes over 500 F to cancel out the molecular mischief in the manganese that always seems to relate to case clusters of Creutzfeld-Jakob or bovine spongiform encephalopathy.

Purdey found intense shockwave bursts, the sweep of Concorde sound, military testing and sound and light and manganese, with deer, cattle and people playing the part of guinea pigs while scientists misread the results.

After the variables were sorted out and put together, Purdey cracked the riddle.

Since 1986, BSE, Creutzfeld-Jakob, CWD (chronic wasting disease) served up a degenerative syndrome. As a consequence thousands of cattle have been sacrificed as if they were propitiation to the gods of Ur.

Moreover, the spectacle has delivered panic to the United States.

While Purdey was assembling his data, Cambridge University biochemist David Brown found that BSE and CJD probably result from exposure of cattle and humans to the "same package of toxic environmental factors; ferromagnetic metals and low frequency sonic shock — and certainly not transmission from one species to another.

Such a finding tampers with the reputation of a Nobel laureate, and therefore invites dismissal.

Purdey's analytical results revealed a high level of mineral manganese, and rock bottom levels of copper, selenium and zinc in all food chains involved. Manganese levels always returned to normal in disease free areas. Animals in TSE (transmissible spongiform encephalopathy) areas always faced front-line exposure to shock bursts of low frequency infra sound, noticed Purdey. Volcanic and earthquake explosions also figured. Low flying super-sonic airplanes created a multi-mile corridor of devastating sound as they landed or took off.

The low copper-high manganese finding rose up like a ghost at each stopping point along the way. "A specific environmental source of manganese could be pinpointed in every TSE cluster," noticed Purdey. There was fallout, industrial dust — always contributed manganese oxide. Volcanic fallout, acid rain, steel, glass, ceramic emissions, lead free petrol refineries, etc., all bagged in their mischief. Manganese, much like silver iodide and aluminum from cloud seeding into the brain via air intake — the old cocaine snorter's route.

The effect of fallout over cattle in pens, with fecal-loaded air forever stirring, can be imagined. High manganese supplements are a fact of life. Spiced-up mineral licks no doubt contribute to CWD in deer. Also, manganese is added to promote antler growth. All species that suffer the spongiform syndrome — deer, human beings, cattle, goats, mink, sheep, 300 animals including household cats — are fed artificial manganese supplements, Purdey's research pointed out.

Manganese is also added to artificial milk powders for calves and human infants — at levels 1,000 times those found in normal fresh cow's milk.

The Food and Drug Administration must surely know that infants do not have developed mechanisms for dealing with such overloads. The resultant uptake of metals in excess of metabolic requirements makes its own suggestion. European cattle make great use of those manganese-loaded powders, hence the precedence of CJD clusters in that clime.

Cattle reared on organic pasture seem to escape the Mad Cow syndrome, albeit not the cleansing rifle shot when bureaucrats go on the rampage. Australia has remained free of the spongiform disease. The manganese supplements are not used down under.

The low copper connection noticed in brucellosis prevention also occurs whenever there is a spongiform cluster. Other transition metals also figure, silver, platinum, lithium, for instance. These metals readily substitute out copper bonds on prion proteins.

Silver mining areas, cloud seeding areas are also implicated in cluster formation.

Inevitably we are led into some consideration of amalgam dental fillings which are still used by many dentists.

The Purdey research deserves a Nobel Prize, not governmental stonewalling. Suffice it to note that low copper, high manganese, shortages of zinc and marked imbalances of magnesium and manganese unlock the conditions that incubate viruses and disease agents. In the case of Mad Cow, the imbalance goes to work directly on to turning the brain into mush without a biological assistant. The bio-electrical impulse transfers via the usual viral pathway, and modern investigators are never happy if they can't find a germ.

Modern agriculture's fascination with the soybean must be noted. The soybean is a manganese and aluminum accumulator the way burdock is an iron accumulator. How much of the manganese madness syndrome suffered by welders and miners is gifted to cattle with errant supplements? Or even by soybean feeds?

On Guam the neurodegenerative diseases run 50 percent higher than the world average, probably a consequence of earlier manganese mines on the island.

Oceanographers tell us that excess manganese is sorted out by the government of this Earth's center of nutritional gravity — and sent to the bottom as layered balls. Just as selenium is a nutrient at four parts per million, manganese is vital, and yet devastating at more than three or four parts per million. There is work, cited by Purdey as part of his investigation, that criminals executed for violent crimes had elevated levels of manganese, "a marker for violence," commented Louis Gottschalk, the researcher.

Purdey's holistic hypothesis has been stated as follows. It was first stated in a paper that proposed that "magnesium could substitute itself at the vacant copper site on the prion protein. This efficient substitution by a foreign material co-partner could induce the all important deformation of the protein that is considered to be crucial to the development of TSE." Mammals dependent on food chains high in manganese and low in copper are therefore at risk.

That, precisely, is what David Brown of Cambridge did. He introduced manganese into copper-depleted cell cultures. Accordingly he produced the key deformation. Earlier organophosphate tests failed to produce the deformity.

Junk food chemists and money hungry entrepreneurs dismiss these findings, even though Creutzfeldt-Jakob victims have exhibited high manganese, low copper in brains at autopsy.

Will this became "settled" science? Probably . . . in 50 years.

In the meantime a cocktail of pollutants will remain commercial fare. Grubs, head lice, organophosphate insecticides are to remain current coin. These penetrate into the central nervous system and exacerbate the more virulent forms of the spongiform encephalopathy.

Just the same, Brown's experiments failed to exhibit multi-replicating properties assigned to any of the spongiform forms. It appears that this transmissible disease is transmissible in the minds of bureau people because it complies with the depopulation remedy, which is invoked for reasons that comply with another agenda. That agenda seems to be implied if not stated in the *Codex Alimentarius.*

The alchemy involved appears to be the same alchemy that attended the efforts at turning base metals into gold. These alchemists considered manganese the black magic metal. It figured in producing a Jekyll & Hyde effect, which may or may not account for the errant behavior that makes the American prison population the greatest in the world on a per capita basis.

Panic, paranoia and head-in-the-sand medicine has long been a USDA worldview, near-Torquemada perfection of the approach being achieved under the administration of veterinarian Linda Detweiler. Torquemada, by way of reference, was the Spanish Inquisitor whose certainty about heaven and hell moved him to torture confessions out of errant Christians and nonbelievers for the good of society. Detweiler is USDA's first line of defense against anything that makes the headlines anywhere in the world — this line of defense usually being annihilation of the herd to confirm suspicion. That the approach is about as effective as drawing a line between smoking and nonsmoking sections in a restaurant, the fact of diffusion of air being overlooked.

After a costly legal battle, sheep belonging to the Three Shepherds Farm in the Mad River Valley of Vermont were seized on March 23, 2001, in response to an outbreak of Mad Cow disease in the U.K. and parts of Europe. Detweiler and associates figured that since the East Freisian milking sheep came from Europe, the probability of prion infection — if that phrase can be used — was great. East Friesian sheep are valuable animals, each worth something in the neighborhood of $6,000.

Biased government enforcers hip-deep in snow trundled the animals aboard a truck for the long ride to Iowa, the killing room and laboratory examination.

One year later, the results still were not in. Any reputable laboratory can have answers in six hours, a fact that Larry Faillace, the owner of Three Shepherds Farm, pointed out. Tissue samples kept on ice for a long period become worthless as far as honest science is concerned, with a so-called false positive the expected readout.

The Iowa facility held on to the tissue for several months, then the materials were farmed out to an Ames, Iowa, laboratory. Two tests were prepared. One was the IHC, and histopathology was also performed. Both tests were completely negative on all animals involved. Had the sheep been cattle rather than "political animals," they would have been pronounced clean.

Sovereign bureaucrats never admit a mistake. Consequently, the samples were retained another six months. In December 2001, one Dr. Rubenstein — who harvested the first results, ran more tests. These also proved negative.

Negative results, however, were politically unacceptable. By tinkering with the protocols and allowing freezer storage to effect its slow deterioration, what appeared to be a positive result was achieved. Even then, in his report to USDA, Rubenstein explained that what he really saw was due to the age of the tissue.

Much later, a report was released that two animals tested positive. No data was released, only press-conference gibberish.

The Faillace legal team asked for the actual data. In compliance with the law, it was released, allowing the facts recorded above to finally be assembled.

It is well known that tissue samples test out of the norm when refrigerator aging becomes a factor. Dr. Bruno Oesch of Prionics AG, the company in Switzerland that does most of the European testing, said "false positives" as soon as the American procedure was explained to him. To avoid incorrect results, he said, he and his colleagues knew they must handle tissue samples right away.

The USDA, confronted with Dr. Oesch's statement, had a spokesperson respond, "We stand behind our test results." This person did not say which test results the "behind standing" referred to.

Through all this, veterinarian Linda Detweiler — who ordered the preemptive strike on the sheep — has kept a discreet silence.

Compensation, required by law, has been forthcoming only in a token way, the Three Shepherds Farm being paid mutton prices for line-bred East Friesian sheep. The 125 animals had a market value of over $6,000 per head. Total compensation has been

$215,000 — less than $2,000 per head, actually just $1,720 per head.

Area growers who surrendered their animals meekly as required by bureaucracy were compensated at the rate of $6,000 to $9,000 per animal. The bureau answer to all this was, "You decided to fight us. That's why you're getting paid less." In other words, go along with the masters of government or be put in your place. The idea of a farmer standing up to his betters cannot be tolerated.

Importation of new sheep, even semen, seems out of the question, according to Faillace. To do so would mean working with the USDA, a difficult if not impossible project. "You can't trust what they tell you," Faillace told me. "You can't make business decisions based on information they give you." When lies stack up like cordwood, correct decisions become impossible, Faillace in effect charged.

The furor over Mad Cow — bovine spongiform encephalopathy or sheep scrapie — has died down, the conventional origin of the mutated prion protein more or less discarded, replaced by the near-certainty that reckless use of Phosmet created the mischief in England and parts of Europe. The veterinary art that asks for the herd to be depopulated to control disease epizootics has graduated into politics, which now attempts to sell the idea of preemptive strikes against political infection.

Background

The term "transmissible spongiform encephalopathy" (TSE) became a vocabulary entry for *Acres U.S.A.* readers in a front-page story entitled "Mad Cow Disease" some 5½ years ago. A mutated protein that punched microscopic holes in the brain had turned up in British cattle, then in human beings as Creutzfeldt-Jakob disease. The almost indestructible agent for this mischief was styled a prion, a still hotly contested hypothesis, cause unknown.

To chase down the cause would require a violation of the holy of holies in the USDA, an indictment of toxic genetic chemistry, specifically organophosphates. It had become the trade practice of university-endorsed husbandry to literally bathe sheep in organophosphates to combat ticks and other insects. Thus assaulted, many sheep rubbed their sides raw before they died, the origin of the word "scrapie." Their carcasses were ground up with pets harvested from city streets and turned into cattlecake, protein bypass, and other euphemistically named feed products. This had the effect of turning herbivores into carnivores.

The agent of infection was seated in the brain, and this was well distributed during the cooking, grinding and powdering process. Surviving disease agents were passed on to succeeding generations.

The U.S. cattle, poultry and swine industries jumped up and down fawn-fashion over the prospect of feeding generations that didn't make it to succeeding generations, as is still the case with poultry and swine. Cattlecake for feedlots was disallowed a couple of years ago.

The cattle industry, smarting under European import bans because of hormone contamination, has crowded the USDA into keeping a lid on the Mad Cow scare. Vermont sheep growers believe they are being unfairly targeted to draw fire based on tests that didn't survive scrutiny, extermination practices that are largely eyewash and ineffective, and targets selected on the basis of "point of origin."

The precedent can hardly escape those who are literate in the subject. First, scrapie is not the same as bovine TSE, according to a British ministry report issued in late October 2000. Second, there have been programs to eradicate *Brucellosis abortus*, Bang's disease, almost since frontier times. Press releases routinely proclaim certain areas Brucellosis-free, usually a consequence of herd annihilations.

The deficits of the feedlot and concentrated animal husbandry are too well known to require exposition. Here it is enough to note that cattle maintained behind feeder bunks loaded with high carbohydrates cannot make it on the best of pastures. In fact Gearld Fry, in *Reproduction and Animal Health*, maintains that cattle have to be bred to thrive on grass. In the feedlot the animal is destroyed by increments, loaded down with acidosis, and probably has *E. coli* migrating throughout body tissue. Unless slaughtered in a timely manner, this animal will die of stress. To suggest that the meat protein such a procedure brings to the market is less than healthy understates the case. The trade calls feedlot beef commodity meat. It has the appearance of food, and with enhancers, it may even taste like food. But the degenerative metabolic diseases that mark the American population suggest an etiology often overlooked, as is most of the cure, a pasture maintained in a biologically correct manner.

18

Pasture-Fed Hog

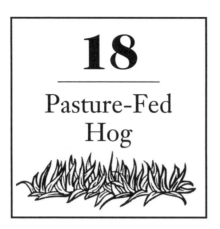

Pasture as a resource for pork production has all but evaporated now that an industrial model has soured the landscape. Massive farrowing operations and pork factories with conditions that are a disgrace to a civilized society have divided communities, destroyed real estate and environment, and brought to market a product about as tasty as cardboard. A small confederation of Cassandras has stayed on to exhibit the form and pasture model for pork production, one without the deficits of America's greatest monument to the stupidity of mankind.

Tom Franzen is a pork producer in northeast Iowa. His pasture-based farrowing operation is a breath of fresh air in an industry that has excused itself from the business of providing quality meat protein as pork.

The point in question has many names, the choice of Franzen being *sustainable*. In growing, sustainable equals pasture farrowing, according to Franzen. Farmers ought to create and sign and possibly hang a miniature of that sign around their necks, figuratively speaking. It should say, "If biological diversity is declining, if topsoil is eroding, if water resources are being depleted, or an increase of external inputs challenges solvency, the system is not sustainable."

"We get trapped by our own thinking," Franzen is wont to say. "The good thing is to back up and decide what a farm really is. A farm is nothing more than a solar harvester. The sun shines on the earth. The earth uses the agency of chlorophyll and photosynthesis to grow something using air, water and a few earth minerals for the purpose of converting energy into a different type of product — grain meat, milk, vegetables, nuts, wood. Then there is a marketing transaction, a harvest of money. Final analysis tells us the only sustainable product comes from the Sun."

Franzen has run the gamut of experiences in hog production. He has constructed confinement facilities and he has settled on a procedure that "makes me feel good," he summarizes. His studied conclusion has been and remains that the "good" is seated in having hogs on pasture.

Pasture means forage, special packages of annuals, peas or rape, for instance, things that can be done other than putting hogs on a flat floor and feeding them computerized feeds.

Different species of livestock can be put together and comingle with hogs on pasture, even having mobile nests for laying hens that live in symbiosis with roving pastured swine.

"We can observe a lot in life just by watching," Franzen says, consciously quoting Yogi Berra. In 25 years his observations have stacked up like well-milled lumber:

- If we have bare soil we get soil erosion.
- If we have uneven pigs in a pen, we get poor performance.
- If I'm next to wind shelter, things look to be better.
- If the hogs are among the trees all the time, things are going in the wrong direction.

Management of feeder pigs all the way to slaughter is the issue. Manure management is also the issue. Finally grass or forage is the issue.

An open operation suggests natural manure management, all in compliance with ecology and economy.

"We do not shell all of our corn. We do not harvest all of our corn. We do not dry all of our corn. We pick almost all of our corn on the ear. Components," says Franzen, "no more, no less."

Ear corn is an option, much as combining is an option, and annuals in the pasture are an option.

In discussing pastured swine, Tom Franzen poses a Socratic question: "What would happen if we would take things like trees, soil, cattle and annual crops and perennials, even interplanted crops and mat-forming crops, and put them all together in a system that makes common sense? Is there a conflict with our values?"

The family farm must first decide what the farm is to look like, and what the values are. The components of an operation cannot be allowed to conflict with decreed values.

That's where pastured hogs come in: use of the resources available in compliance with real values.

Starting in 1991, Franzen has invoked these stated principles. He grows a narrow strip of corn, followed by the same-sized strip of small grains, followed by a strip of legumes. He pasture farrows on a strip of legumes. Legumes then are rotated back into corn.

There is that old saying in *Paint Your Wagon*, "There has to be gold in them thar hills, elsewise where did the idea come from?" Franzen might chuckle at this yarn. In fact, the idea that has taken hold on this Iowa farm came from an old man who dropped off this advice. "If you want to do something smart, you'll learn to harvest annual crops directly with livestock and put it into rotation." The old timer said dung and urine strength on the soil is ecologically the soundest thing you can do to produce the highest yield in annual crops.

The ability to accept good advice is rare, about the same as the population that buys quality rather than cheap price. Resultant activity from that conversation was the growing of annual crops. Clover, a rotation back to an annual such as corn, small grains, then a legume.

This sequence established, Franzen next added components. The strip came to include corn, oats, red clover or alfalfa, or a combination, with a shelter belt for trees in between. There are different types of trees, red oak, green ash, poplar, cottonwood for quick shelter, nut-bearing ash, hazel bushes — a mix planted at distances sufficient to allow drain tiles to function.

Norway spruce on the north edge of a field running north and south are a part of the general layout. "We want long term hardwood use, short term wind shelter and a fairly short term economic use out of nut-bearing feeds.

So-called conventional farmers ask, how can you take good Chickasaw County, Iowa, rich black fertile soil and remove strips of land out of production of row crops? The answer is "harmony of values."

Any discussion of hogs on grass without this preface would be quite beside the point. Taking a combine out of the system complies with real ecological values. Not to combine harvesting corn seems anathema to farmers who slavishly follow the industrial model. Hogs in fact harvest the corn with no payment involved, certainly without the cost of diesel fuel and machine depreciation. Moreover, hauling, drying, unloading, the transferring, the feeding, the whole nine yards of energy-consuming activity.

Franzen computes his benefits at $50 an acre when he lets the hogs do the work. Others can be invited to make their own computations.

If a 50-foot-wide strip can save $50, it can effectively pay the farm entrepreneur "rent" for itself, as economically implied by the narrow strip of hardwood. Internal savings capitalize the general rejection of the monoculture concept.

This modest family farm fed on grass and diversity can afford an underground water system. Well water, T-ed at junctions, and running three-quarters of a mile, cranks out an answer to drought, seasonal burnout of grass, providing a comfortable run for hogs on grass.

On the grass-based farm the water cycle and the mineral cycle complement the grass base for pastured pork.

Holding expenses in tow is greatly dependent on judicious handling of the mineral cycle. "We produce our own protein, our own energy and we harvest that energy with our own animals," the hogs on grass farmer reported.

Looking back, 1992 was *the* test. There were 37 inches of rainfall in June and July. The scourge of that much weather would be devastating for most farms with free ranging hogs. Here it was not a problem, with hogs ranging in separate groups with not over four to six sows in a group, each group separated by a strip of corn and a strip of small grains with a belt of trees adding to the mix.

"We had actions compatible with values. Strips were about 600 feet long. Those red oak trees take 50 years to "grow up," as the children put it. One of the children observed to her dad, "You're not planting them for yourself."

That, precisely, is the role of the family farm and its values, and in a sense its devotion to grass, free range for animals, and the future of the erosion-free land once the present generation is gone.

Draining away excess water, controlling animal migration with electric fences — again, these are operational systems compatible with family farm values. Tom Franzen does not want his children exposed to chemicals of organic synthesis, for which reason those materials clash with family farm values.

Moving fences and performing chores that are safe and valuable involves children with compatible values, Franzen has pointed out.

The electrical system is fed by underground wires so that any section can be energized without a long trip to the feed. The underground system for water is protected so hogs cannot root down and disturb it.

Then how does the hog on grass perform? "It turns on its own lights, it opens its own door, it hauls its own manure, it does its own disinfecting." The electric fence, four or five inches off the ground, permits tire traffic with vehicles without stopping. Hog huts are moved with a high loader, all from the seat of the tractor. The hut, in short is moved without a human hand touching it.

The bottom line is one that permits bred sows to harvest their own protein, spread their own manure fertility, under conditions of low stress, with a tight age requirement being maintained when they have their pigs. Even a death loss is less because the investment is less.

This particular farm offers a lesson that cannot be ignored with impunity. When the initial investment is less and maintenance expenses are less, the profit goes up.

This is not a system with a lot of runoff. It is a system that develops a diversified root system that keeps the soil surface covered and manages water penetrating into the soil profile.

It is management of the mineral cycle that most offends values when crops are removed from the farm. Animal harvest of protein in corn and grasses and small grains retains for soil use the very minerals that leave in the beds of transport vehicles. Manure returned to the soil handles those minerals and improves soil fertility. This manure revenue complies with family farm values and makes money at the same time. The sow herd needs little outside feeding when it is allowed to pasture. Except when there is snow cover, grazing is the operating ticket under a forage program.

"I maintain that grazing a sow herd for just two weeks in the fall will turn enough profit to buy all the fencing supplies for a whole year," Franzen summarizes.

The general objective on the farm that backbones this chapter is 100 percent retention of nutrients from runoff. Picture runoff as a drain out of the bank account. The final question is, simply, is this system cost effective? Does it account for a return for dollars invested? Franzen has his life and capital invested. Roll the bookkeeping over a season, everything measured, revealed that the return has served up several lessons. The school book professor will tell all who listen that pasture maintenance of sows and attendant farrowing is labor intensive. Labor intensive compared to what? Are we to believe the modern hog factory inside is not labor intensive?

At 7 a.m. Franzen takes care of six units of pigs. Eggs are picked up at 7:15. In 15 minutes he can take care of six groups, typically once a day. All hours logged for a year it came to an hour of labor per litter of pigs.

If "labor intensive" as a norm is demolished by an actual record, then what about productivity on grass? Sows gestating on grass, then weighed at various stages, reveal that the cost of producing a feeder pig was cut in half, this compared to the industry standard.

In a manner of speaking, sows seem more intelligent than a lot of farmers. On a blustery day, sows know enough to find shelter behind a corn stand. Hogs know the wind blast in northern Iowa is raised off the surface by the intermediary shelter, whatever its type. In hot weather the grazing hog finds shade. Little pigs like to play in the standing corn.

Hog huts are kept in a row, and tractor traffic stays in the same tracks.

Come September, the corn is mature, the electric fence is pulled away. Gilts head straight for the corn. The ability to move fences with relative ease makes the system work. Once weaned, piglets are moved into hoop buildings for finishing. If there is ground exposure, it will be in a fairly narrow strip.

Corn loss? Close to zero, Franzen's records reveal.

The nuances that attend farrowing and finishing would require a manual the size of this book if fleshed out in detail. The point here is pasture and pasture's contribution to the profit profile in a world of technology that has discarded the grass connection in pork production. The ability to dismiss overhead expenses — the type that ought to be furrowed out, defines the real values of this family farm. The four wheel drive vehicle that is used profitably only occasionally is an example, as is superfluous green equipment, especially the corn picker.

Hogs like sanitary conditions. They like forage. They definitely like air that is unpolluted by ammonia. Most of all, they like to drop their manure and move on. Sows are in corn only in September and October. The rest of the time they rely on forage and such supplemental feedings as conditions make necessary.

"Don't fight nature" is more than an adage. Now it's back to farrowing with the seasons, Franzen says. Farrowing in March in hog buildings has replaced an element of the conventional ignorance, meaning farrowing all year round.

Pasture farrowing can take place only in May, June, July, August, September and October, this in Iowa. October is red flag time also.

Don't fight the seasons, is a parting shot. Hog houses are made of fabric and have limestone floors. The product litter-wise can be one to two pigs less and still deliver profit in tune with family farm values. The goal is profit per acre, not records for silver cups or industry citations.

Genetics — Hampshire, Duroc, York, etc., — always figure, but not as much as the grass connection.

Withal, grass has made the electric fence mandatory, and the electric fence has made the hog pasture a reality. More important, both of the above have made pork production on the family farm a working reality, this at a time when academia and Extension have both blessed manure hog concentration operations that foul entire counties.

There are nuances a'plenty that invite attention, judgment, and solutions that may or may not recall the traditions of yesteryear.

As far as the one hog producer discussed here is concerned, the jury is still out. On the business of ringing the noses of hogs, "It may not be necessary," Tom Franzen told a group of assembled eco-farmers. "If the animals are destroying the grass cover, it may be indicated." As a matter of family practice for father and son, "I still do it," he said, but he trailed off by expressing his doubt,

doubt being ratified by perhaps 25 percent of the sows being without rings. The more room a sow has, the less likely she is to dig.

Bedding is a yes-and-no proposition. Cold weather asks for bedding, hot weather — "No!"

Franzen summarizes. "To wean, I back up the trailer with corn aboard. The hungry pigs pile aboard. I close the gate and we're on our way. If they refuse to climb aboard, they'll be hungrier tomorrow."

Bad weather is as good as good weather when it comes to weaning.

Bottom line: get the mixer and feed grinder out of the system, forage and corn strips in. As Wendell Berry says, "Expensive solutions to agricultural problems are almost always wrong."

Shell corn proves the observation. Sows first tear the corn off the cob. Then they polish up the kernels. Finally, they eat the cobs. Efficiency is individual. Growth is uniform. No sow becomes bred when feeding on ear corn in addition to forage. The full inventory of feeding practices constructs a venerable raw-food feeding book, minerals included, but the gist of the system remains room, air, forage, corn and a rotation of feeding days for the several breakfasts of feed, ear corn four days of the week, oats, minerals, etc., spaced between.

19

Poultry Pastures

Earlier in this text, the point was made — not too subtly — that grass is a bit more than the forgiveness of nature. It is the broad spectrum support for the diversified farm.

Understanding any livestock management system is the approach to soil stewardship. The very term pastured poultry brings academia up fighting from its chairs. Not only has confinement poultry been ratified, the idea that birds can benefit from anything except grain has been dubbed settled science by the Tysons, the few dozen "growers" who hold their participating farmers in indentured servitude.

Over the last dozen or more years, a few innovators have stepped forward to recapture the values, and those values were found in residence on pasture.

Tom Delehanty of New Mexico is one of the founding members of the American Pastured Poultry Producers Association. He operates the largest pastured poultry farm in the country, producing some 50,000 birds a year. The nuances that attend pasture maintenance come together in the soil system itself, a fact often overlooked by farmers west of the Isohyet. Delehanty is quick to credit Joel Salatin for the insight his model has conferred on all those who saw poultry on pasture as an innovative side departure from the confinement system.

Delehanty came by some parts of his farm view quite naturally. He grew up on a Wisconsin farm milking cows, feeding pigs, handling chickens and turkeys — "a veritable Old McDonald's farm," Delehanty now recalls.

The poultry end of farming challenged Tom Delehanty. For reasons that eluded him, he found chickens too lethargic to go out and forage. It was the idea of moveable cages that caught his attention. Dale Schurter of the old Ambassador College built such cages before the 1970s, but the idea didn't move out until well after eco-farming got underway.

Salatin's books and appearances sparked a movement, and the movement folded its security blanket over the life and goals of Tom Delehanty. He moved to New Mexico in order to escape the six month restriction climate and weather imposed on him in Wisconsin. The idea of moving cages by the numbers suggested itself. The cages could be moved every day, even twice a day, the birds would eat the bugs, the tender greens, the seeds forage had to offer, all the while perfecting the soil in a powerful way. A year-round small scale production model could support itself.

Tom Delehanty can recite benefits generated by poultry on grass with calm aplomb, always reminding his listeners that the soil benefits most.

Delehanty put it this way. "We live in a desert. Rainfall is 8 to 10 inches. We water once a year. The neighbors water ten and twelve times a year using flood irrigation."

Using his water conservation system, "We're bright green year around. If I put a crop in the soil we graze it through the next summer. Thus the need to irrigate in the winter is erased," Tom Delehanty summarizes.

The intercropping between vegetables and poultry recommends itself. Instead of using a lot of equipment, the pasture poultry grazer uses people, eight to 10 employees, this a way of helping the community in which employment is hard to come by. The bottom line for employees and the farmer is a supply of healthy chickens, this even without city community gardens for employee use.

The soils are heavy clay. In the dry season, most of the area soils are hard as a rock. Squeezed into bricks, the soil is suitable for adobe construction. The area irrigation system silts over creating a hardpan. Mechanical ripping and biological inoculation with poultry litter heals the soil. After three years of tillage and poultry traffic, the worms show up. "I think what the birds offer the soil is the microbial exchange. The birds are eating the soil. This fact combined with mineral correction ends up sheet manuring the field."

The New Mexico operation has some 120 pens "on the move." Tom Delehanty sees his model as sustainable, one that can support a farm. The question is grass.

Research has proved that fine cheatgrass can support herbivores such as guinea pigs, Canadian geese, various species of foul — yes, even chickens. Actually, chickens find plenty of seeds in the pasture. More to the point, they have access to grass before it joints, which is precisely when almost 100 percent of the nutrients are available.

The Rio Grande part of New Mexico is nearly dream perfect for year-round poultry production.

Some 16 to 20 brooders service the operation. Gas heaters with portable gas tanks serve up winter heat. An even 1,000 birds are housed in each starter unit. Three weeks is enough time — sometimes four weeks — is enough brooder time. Brooders are moved to fresh grass each time a new batch is started. Their comfort is assured by shaving bedding. Baby chicks start eating tender grass within two weeks.

Organic grain is an important component in the feeding program. For several years, Tom drove 1,000 miles a week simply to secure grains that were not genetically modified. This shortfall has repaired itself as a state supported organic farming program gathers speed and moves ahead.

Good feed, feed without chemical residues, is ever a challenge that even good pasture won't cure. On scene grinders seem to be a must, as are connections with grain growers who take seriously the spirit and passion of certified organics. Delehanty operates with a backup system. If a grinder breaks down, a spare is quickly pressed into service. If both break, the older one is robbed for parts.

There's more to pastured poultry than brooder houses, moving cages, bins and grinders. The mix itself becomes an art, albeit not the only consideration. It is the role of grass and open-air living that seems to confer health, finally taste, to the end product.

The entire package comes together for Tom Delehanty and family with the use of old machinery, thrifty investment and family work.

Winter wheat and winter rye are planted in late fall. The key to grazing is to have the birds take the tender green shoots during the brief season sprouts are available. Birds relish seeds, insects, and they positively cherish the protein heavy grasshopper. By the time serious grazing gets underway, there will be 60 to 65 birds in a pen. Each pen is scrutinized and managed each day. "If they seem to not be eating their greens we pull back on their feed," Delehanty points out. The health of the bird requires that pasture salad. Birds can and do overeat. In fact the industrial poultry business is designed to bring on overeating. The lights are kept burning through the night with the devilish intent of causing over-consumption. Digestion problems prompt drug sales and also cause deaths. Great death losses occur if the floor isn't cleared, and all birds slaughtered are a day or two too late.

Pastured poultry growers blanch when reminded of trade practices in Arkansas and Alabama and other growing areas. Dead birds as well as cats and dogs harvested off metro streets, end up as refined "feed" for chickens and hogs, all with the blessing of government veterinary and feedstuff experts. The pastured poultry grower has none of this, of course.

"You have to choreograph your fields," according to this New Mexico pastured poultry grower. "We pick up pens, stack them on a truck trailer, move to a new field and start over." A lot of organic material is left on the ground with each move. The means of water, food and equipment management are on par in importance with the management of grass itself. Water has to move with the birds, and this is accomplished via water transport through quarter inch feeder tubes.

With 1,000 chicks arriving each week, the shape of pasture requires advance planning.

Chick Life

The real life quality of a growing chick was examined long before the industrial model pulled the family farmer's mortgage lifter into moveable poultry houses.

Healthy feed fare cancels out the need for antibiotics and hormones in feeds and expensive coccidiostat remedies. The medicinal quality of young grass was examined in depth by Charles Schnabel, as was noted earlier in this text. Suffice it to say the basics of those findings speak for themselves.

There are warm temperature grasses and there are cool weather grasses, winter grasses and summer grasses.

About 50 F is a good daytime average, 28 F is an average low for December and January in Soccoro, New Mexico. Delehanty's farm is right off the Rio Grande as it crosses New Mexico. The farm is close enough to town to enable the use of city water for processing.

There are times when snow visits a foot high blanket on the scenic landscape. Straw becomes the poultry grower's friend when grass is held in escrow by a defining nature. Straw with heads on it invites birds to do what birds do, seeking grain. Pads that keep the birds off the turf during snow and wet weather become a population dense colony of worms.

Adobe soil ceases to be adobe as cages cross and recross fields. The spectacle of workers digging worms in the middle of the summer has agronomists taking a new look at soil in the harsh semiarid New Mexico outback.

It takes about two years to reclaim soil when the chicken biology connection is reestablished.

Withal, it must be stated that the pasture element is merely a single factor on most diversified farms. Innovation has put pastured poultry and eco-correct eggs back into the market, with the movement growing year after year. The deficits for such operations have become books, compendiums and a bit more than organic folklore. As entries at farmers' markets, the pastured poultry continue to introduce taste into poultry consumption while nibbling away at the oligopolies that appear to hold the trade in thrall.

Consumers want fresh chickens, eggs, lambs, beef, pork and the pasture label confers comfort in advance of consumption.

Behind the farmers' market are systems too numerous to mention. One and all, they are based on simple, durable, tough shelters that can be moved each day with the power of the soil released for regeneration. Soils, ever a curse, that can develop saline seeps seem to bow to that microbial invasion pastured poultry empow-

The Warning

Few people realize how near its harvest has come to failing, and as result how close the human race is to nutritional disaster. The prophet Ingalls was not talking about the world's 25-billion bushel grain crop or 20-million ton sugar crop supplied annually by the grass family. There was a time when these two grass harvests did not exist. Man's debt to grass is a hundred times greater than that.

The basic unwritten theme of human history has always been the search or struggle for greener grass. It is no accident that the highest civilizations of the past have coincided with the best grass land, and these civilizations have failed largely because of man's neglect and abuse of that grass. Most of the cereal grasses and human life apparently originated in the same part of the world. Herodotus, the Greek historian, described the "fertile crescent" as a land of *unbelievable fertility* as late as 450 B.C., but much of this area is now a desert.

Transportation has postponed the doom of the present civilization by continually opening up virgin grass lands, but we are now at the end of the road on that score. We must grow better grass or perish.

America and Americans are what they are, largely because our boundaries include some of the best grass land in the world. It is certainly no credit to American agriculture, however, that the grass plains once supported more buffalo per square mile, and in much greater vigor and fecundity, than the same area now supports "animals units."

All the accumulated knowledge of medical and nutritional science can't suggest a dry lot ration that will duplicate what young grass does for the health of our winter-starved animals when they are turned out to graze in the spring. Yet in spite of this annual miracle in animal nutrition all text books on feed and feeding formerly taught that young grass was "washy and without substance." Because of this fallacy farmers for centuries have shamefully neglected and abused their grass pastures. Many farmers, even today, admit that they "never put a field to grass until it won't grow anything else," and think this is the proper thing to do.

If all grass is eliminated from a poultry ration only 20 percent of the pullets placed in the laying house in the fall are fit to carry over a second year. This means an average productive life of only about six months, yet a hen is potentially capable of laying 4,000 eggs and will continue to lay for twenty years if she gets the right feed.

The "secret of grass" was discovered in trying to find the cause of the tragic failure of a large hennery due to disease after the pullets had been ranged on the prettiest stand of alfalfa any poultryman could ask for. The productive life of hens was *quadrupled* after they were given all the 30 percent to 40 percent protein grass they could eat for eight months. As the amazing story of the nutritional value of 40 percent protein grass unfolds there are increasing grounds for believing that the reputed age of Methusalah was not fiction but recorded truth. If some favored spot in the world once produced 40 percent protein grass in continued abundance a Methuselah was possible.

The secret of grass explains the peak of milk and egg production in the spring. The summer slump of milk and egg production is not due to hot weather, but to the fact that most of the grasses have jointed and matured, though there may be young weeds and legumes in abundance. There is no doubt the spring time rate of milk and egg production could be maintained the year around if cows and chickens got only 1 percent of their calories from 30 percent protein grass.

er. The pH in the New Mexico Rio Grande Valley is often a high 8. The poultry fix has reduced the pH, re-established organic matter.

It takes about 30 acres to handle 1,000 birds a week using the pasture model.

The new unidentified trench among the new farmers emphasizes the pasture, but diversity counsels moves in many directions. "I bought myself a 40 year old packer-sheller. I am going to get into sweet corn because it is an easy crop without major labor. I expect to grow open pollinated varieties and saving some corn, trade, barter. Garlic works well in New Mexico, potatoes have my interest. You can plant in February, harvest by May," Delehanty said.

As a parting shot, Delehanty has to admit he mows four and five times a year. Poultry enrichment of the soil puts to bed the argument that drought closes down growth.

Depson

Depopulation has now become a favored federal veterinary practice. It proposes to eradicate a disease in an area by killing all the animals, even those that exhibit no measurable reaction.

Murray Bast of Wellesley, Ontario, has documented the efficacy of using hydrogen peroxide in the drinking water of poultry, pastured or otherwise, with excellent results in controlling avian diseases.

Probably the best work on avian diseases is Robert Stroud's *Diseases of Birds*. Stroud became known as the Birdman of Alcatraz. Sentenced to life in prison for a felony murder, he became a model prisoner, finally a self-educated expert on bird diseases. Often in solitary confinement, he acquired endless patience. To pass the time he tamed sparrows that flew into his cell.

The warden became so impressed with Stroud, he agreed to allow his prisoner to breed canaries. He became well known among canary breeders in the U.S., even worldwide. Calls into the prison for advice became the norm.

When Stroud's birds came down with avian flu (he called it bird diphtheria), he found no established remedy. He searched the medical literature, veterinary and human, and learned about sodium perborate, which he could use to sterilize the water. This cured his birds of avian flu.

This was the singlet oxygen discussed in Chapter 17, the atom that killed the infection. It was the effect delivered by adding Proxitane to the drinking water of birds and animals suffering a variety of diseases.

Stroud found that the formulation did not retain its medicinal benefits very long. As with hydrogen peroxide, a powerful oxidizing agent, it worked rapidly. Peroxygens react vigorously with contaminants, oxidizing them. Stroud found a stabilizing agent for the O_1 factor, a finding that put remedy within easy reach of every poultry grower. He hoped to earn enough to keep his aged mother once the stabilized remedy became commercialized.

It didn't happen, of course. Prisoner Stroud's calls were restricted and he was restricted to two letters a year, these to his brother. It was pointed out to me by England's Alwyne Pilsworth that this sequence of events was happening at the same time William F. Koch fled the United States with his cure for cancer. Koch was using ozone, probably the most effective peroxygen available. In addition to providing the O_1 singlet oxygenation, it also supplies pure oxygen O_2.

Hydrogen peroxide is a good example of the sensitivity of peroxygens to contaminants. H_2O_2 is so sensitive it takes only a dedicated sprayer to serve the farmer who uses H_2O_2 as a growth promoter. The full story of hydrogen peroxide can be found in *The Therapeutical Applications of Hydrozone and Glycozone* by Charles Marchand, republished by Water Grotz. The original publication was 1904.

H_2O_2 is much cheaper than peroxytane and carries 35 percent H_2O_2, as compared to 20-percent-H_2O_2 peroxytane. However, peroxytane is easier to use as a sterilizing agent than H_2O_2 because the formulation is more stable. Peroxytane remains stable for up to 24 hours, all this depending on the degree of contamination in the water.

Peroxytane also contains a wetting agent to improve performance. H_2O_2 has to be metered as used or served up in distilled or de-ionized water.

Blood-building material in a poultry ration is a limiting factor in egg production. Further, a lack of these materials is even a cause of mortality.

This reality brings into focus the chemical and functional relationship between chlorophyll and hemoglobin, making the green leaf the best source of blood.

Pasture for poultry answers these considerations. Some growers attempt to deal with this requirement by increasing the alfalfa meal component from 5 to, say, 20 percent or 25 percent. According to Schnabel, anything in excess of 10 percent is actually harmful because of kidney stress.

The diuretic principle of alfalfa is seated in the leaves, 15 percent protein for grass can be bad at higher levels than alfalfa with 30 percent protein.

Alfalfa has a sort of folklore status. Schnabel not only questioned, he substituted greens at 5 percent and 10 percent one at a time. He also fed the common vegetables one at a time, five to 20 percent of the ration. After a three-year search the hunt for answers settled on high protein wheat and rye as grasses. Hemoglobin regeneration became a fact. The do and don't aspects filled journal after journal. Dried spinach was found to be harmful because it interferes with mineral metabolism. Tannins, alkaloids, saponins, all inhibit the sought-for process.

Researcher Charles Schnabel was about to abandon the project when wheat and oats sprouted as a grass made its suggestion. He put chopped grasses into the ration, simulating birds on a free range. As the season progressed, sprouted oats replaced the greens.

As compared to the controls, the greens-fed flock averaged 94 percent production for five months, May through September, 98 percent for June and July. Controls declined from 45 percent to 32 percent.

Equally important, the greens-fed birds were free of degenerative diseases, the usual consequence of forced production. Livers achieved a dark mahogany color. Combs kept their scarlet red. Leg pigments never faded.

Dogs that chew grass seem to doctor themselves with the pasture diet they seek in place of protein-laced fare. Similar results have been codified from laboratory animal experiments.

On balance, pastured poultry harvest health from their exposure to sprouts in the field.

There are a few caveats that attend pasture blessings.

1. Dehydrated grasses must be 30 percent protein grasses. Lower quality grass will not do the job.

2. Grasses must be available to poultry just before they joint.

3. Vitamins cannot be preserved unless any drying is accomplished quickly.

4. High-protein grasses at 20 percent of the ration are indicated.

5. Feeding more than 3 percent meat scrap is inherently dangerous.

6. Damaged livers do not fully recover.

7. The pasture can only serve as an add-on, not as a complete feeding program. This suggests a sprouting program for tender grass production on a routine basis through the year.

8. Perennial grasses seldom contain over 20 percent protein.

9. Few birds can find a full complement of grass on range consistently.

Thus are stated the benefits and some of the shortfalls of poultry on grass. Doubtless some few growers will take these facts by the nape of the neck and seat of the pants and shake out new results in the fullness of time.

20

A View from South Africa

The lessons that have come to me in this quest of pasture knowledge have stitched the world together beyond the wildest dreams of traders. One lesson cherished by many readers was delivered by John Fair of South Africa, with special attention to the High Veldt. On the map, Fair talked about the "Orange Free State" — a political unit with no oranges, no freedom, and in a terrible state. His area as an Extension officer has diversity, yet the common denominator is pastures.

Much as is the case with American farmers west of the Isohyet, the High Veldt depends on favorable weather for the forage that ample inventories of sheep, cattle and other grazing animals require.

Even with good weather and ample grass, the debt load of farmers tormented the grower's bottom line in almost every area encountered by the diversity of farmers in South Africa's own high plains. Like it or not, all the world's farmers seem locked into a race for the floor.

Yet fully 30 percent of the land in Fair's area was composed of good, deep soil that could hold its moisture. The average farmer saw his profits on such lands. As with most U.S. farmers, the "Orange Free State" growers all seemed to have land with lower potential, land that should never have been plowed. Those lands ate up all the profits. It was that paradigm that needed correction.

Folklore dogged Fair's farmers no less than they govern most farmers the world over. When it was suggested that "humus" meant more than soil depth, the paroxysm of laughter hardly subsided. With or without advice, most farmers seemed satisfied to plant corn or some other favored crop on marginal ground the same as on that usual 30 percent best calculated to deliver a yield.

As an intellectual adviser to area farmers, Fair counseled taking marginal lands out of row crops and returning them to pasture. In America the fiction has long survived that "when this ground won't make corn anymore, I'll put it to pasture." A fragment of that thinking seems to be a part of every culture.

In illustrating the translation to forage, John Fair could point to a special grass, *Agrostis thurberiana*, which required expensive machinery to translate it into feed. In terms of cosmetics, the feed looked good, but "it proved to be a hopeless system," according to Fair. Expenses took care of all the profits, leaving nothing for the bottom line. The cost of feeding that grass proved to be prohibitive.

Growing forage on worn-out land calls for the nitrogen that absence of a natural nitrogen cycle requires. Purchased nitrogen was delivered for two or three months, Fair reported, but the run of satisfactory production ended right there. A local producer called for a burn-down of residue. This enabled the farmer to live off the ashes, but this approach was far from sustainable.

To combine one foolish idea with another, local practices combine *Agrostis* with lucerne. The genus and species proved incompatible in growth and soil requirements. Grasses will grow almost everywhere, the quality being dependent on pH, nitrogen, calcium, potassium, phosphate and trace elements. If available, many grass species uptake more trace nutrients than any other crop, tree or plant, whether characterized as food, weed or legume. Lucerne is very choosey about calcium levels.

The next grass that asked for attention was kikuyu. This grass was an import from Kenya. A big-game hunter found it growing around the kraals of the Masai. The grass was excellent for rugby fields. On the farm the grass fell victim to a perpetuated lie at the university level. The special falsehood, cemented into place by the university, was that kikuyu would grow only with an annual rainfall of 800 mm. Yet farmers using the species as erosion control on compacted soil in an area where violent rains and long, dry periods prevail have said that it grows quite well. In fact, the John Fair observation is that kikuyu will grow at "less than 300 mm rainfall." *Grow* here implies that the grass gets the right nutrition. Here was a grass that would grow on a very acidic soil, but it had a voracious appetite for carbon.

The key to this particular grass is the same one many American farmers have discovered for forage preference on both sides of the 30-inch-rainfall Isohyet. The trick is to get all the manure onto the pasture in winter. In summer it becomes a low-cost pasture-fertility load. By removing feeding bins and utilizing rather than disposing of manure, the forage gains permission for life. For a quick start, rows with concentrated manure and fertilizer jump-started the grass, lush growth increasing incrementally in two to four months.

Nitrogen inserted into kikuyu pasture cuts like the edge of a knife. It has to be applied "on the plant" became a John Fair finding — this to get the plants off to a good start.

The next grass that invited attention was Smuts finger grass. As a sheep pasture it humbled the best of operators. "You have to be a master to make it work," was Fair's conclusion. If there is a salvage point it is that this species will grow on an acidic soil.

Mentioned so far have been acid-tolerant plants. Fescues come next. The suggestion presented itself that in the mild climate of the veldt fescue could be grown in autumn, then used as winter forage. The root system sucked up nutrients on which sheep and lambs thrived. The average gain of a pound a day during winter rates attention as being quite satisfactory.

Shortening the winter is always an objective. Fescue, however, has a hard time surviving hot weather, is slow to establish itself as a permanent pasture, and rarely migrates via natural selection into permanent pasture. Still, survival happens, sometimes in competition with strong grasses. Observation, site selection and use of harvested seeds nevertheless established some 60,000 hectares (about 120,000 acres) of fescue planted during a given year in South Africa. The conclusion was "very robust, very tough, a good performer."

In trials, ewes continued to gain en masse, an exceptional result. Growth of lambs came to a half-pound a day.

Pasture legumes often provide a better option, such as the annual serradella, which Fair calls "Cinderella Serradella." Its slow start is forgotten when the finished pasture materializes. Planting between rows of corn permits winter grazing by sheep and delivers wonderful lambs. At the end of winter the serradella is still there. This ideal crop system features crop and grazing, replanting extinguishing the forage, then permitting replanting. Bringing livestock and a crop together represents an innovation often utilized in the United States when row-crop acres are being returned to pasture, albeit as a first step generally. On the South African scene, triticale was planted as a subsequent crop, the benefits of this nitrogen-fixing plant being evident as the new crop matured.

Fair could point to nitrogen fixation as the reason that oats would grow where they hadn't been grown before.

The exchange of plants internationally has been a fact of life since before the time of Thomas Jefferson. Yellow serradella from New South Wales is persistent, always reseeding itself as it does in Australia.

The saga of vetch between rows of corn is equally instructive. Much the same has proved true for various legumes, clover included. There is a relationship between clover and bloat. Some farms plant clover with no resultant bloat, while on other farms such pasture seems to be absolute poison. The cation exchange capacity governs, and errant use of forages regardless of soil composition, nutrient load, moisture or a carbon complex can deliver a plant in the fullness of time, but animals can starve in the midst of plenty, as they often do in Florida and the rainbelt South, where they graze on lush-looking but impoverished grasses.

Of interest has been the results Fair and his farmers were able to achieve with alternative legume pastures. Shallow-rooted plants require irrigation scheduling that is very near perfect. Any faltering in the management chore puts a loss of production on the other side of the equals sign. Any deviation with nitrogen fertilization also equals loss of production. A routine harvest of, say, 16 metric tonnes of dry material can easily drop to 12 tonnes. On average, the Fair record revealed 1,900 pounds net gain per acre, this figure reduced to 1,400 pounds live net gain per acre. In other words, the legume gave far greater live net gain of salable material per hectare. The cost was less by far.

Much the same was true with perennial clovers. One case report should suffice. Fair cites 2,000 live net gain, nearing the gain of the animals during the months they grazed on the forage.

Pastures identify the soils on which they are grown. High humus, high fertility of bottom lands stamp their imprimatur on the forage they produce. The years of maximum production extend themselves according to the environmental capital that nature has bestowed. Looking back, John Fair divides his career into two periods: B.K. and A.K. — before and after Neal Kinsey. It was Kinsey who instilled a new appreciation of the Albrecht system in South Africa, and this includes the pasture lessons contained in the chapter on Kinsey's tryst with pastures.

The bottom line seems to be one constructed on a base of low-cost pastures. Fair has been able to calculate the cost of producing milk at one to three cents per liter when the client relied on forage as mixed by nature and as fertilized by manures freely distributed by the animals themselves. One fellow "hasn't spent any money on his pasture in 15 to 20 years."

Crown vetch is an heirloom variety worldwide. Incredibly resistant and hardy, drought tolerant, almost eternal. In areas where it was planted for erosion control 18 years ago, it still flourishes. Even a fire cannot destroy it, since the ashes furnish fertility for a jumpstart to new life. As a pasture for dairy, it recommends itself and awards its own prizes. Crown vetch does not care about political boundaries or even most environments. Unfortunately, it cannot deliver trace nutrients that are not available. It can, however, crowd out unwanted pasture invaders sometimes styled "weeds."

Pastures, it turns out, are not merely cover for throw-away acres. Their quality demands the full spectrum of major, minor and trace nutrients. With the proper fertility load, the grasses, legumes, herbs and even weeds construct their own hormone and enzyme systems, ward off root diseases and nematodes, and construct battle formations capable of rejecting bacterial, fungal, viral and insect attacks.

Failure always suggests a soil structure run out of nutrients, or one having complexed and confused or locked up those in inventory.

Nature has never engineered a plant to withstand starvation, even if their existence suggests as much. There is a variable that is hard to calibrate. The revenue of energy from the sun is assumed. Not so clear is just what traces plants take from the air.

Lespedeza is "coming out tops" in South Africa, according to Fair. It will grow on the worst soil, on sorry soil, if only it can survive the first few years. Lespedeza is taking command of hundreds of thousands of hectares. Unfertilized, the nutrient quality seems to be not much better than bedding hay, thus the genetic engineer's dream of a plant that can live on nothing yet deliver nutrient value seems to be no more than the dream of avarice revisited.

In the A.K. period, Fair expects to see the pasture scene change dramatically because the little professor at the University of Missouri taught a man named Kinsey, who now roams the globe teaching and spreading the Albrecht message well beyond the exposure Albrecht was able to achieve in his lifetime.

Lessons transplant themselves, for which reason I now summarize an incident recited by Fair at an Acres U.S.A. Conference.

He consulted at an area known as the Mecca of South Africa. It is the coastal belt, the pasture belt, and is about as close to New Zealand as you can get.

There he met a veterinarian named Dr. Alfred Kit, who said, "John, we're sitting on a time bomb. I'm monitoring herd health. The potassium levels of the blood are three to four times higher than they should be. Blood urea levels are high. The first symptom of this problem is fertility. This problem was and is the failure to conceive, which has to do with the inoculation of the uterus of the calving animal with high potassium. The uterus doesn't contract quickly after calving and expels the remnant fetal fluid, so you have metritis."

This problem is on the increase. Consequently, advisers told farmers to slow the application of potassium to pastures. Pasture productivity dropped. This was answered with more nitrogen. Nitrogen and potash increased in the herbage, K moving from 1.4 to 3 percent.

Poor pastures mean poor cattle. Problems with feet followed the nitrogen-potash uptake, calves suffering from osteoporosis. This syndrome is explained in Voisin's *Soil, Grass and Cancer.*

There is a staggers index equal to calcium divided by potassium plus magnesium, each divided by the milk equivalent. The numbers Fair came up with were 4 percent, 22 percent, 0.28 percent. Plugged into the formula, the result is about 2.7 percent. At less than 1.8 percent, the pasture manager is safe. From 1.8 to 2.2 percent, there is danger. More than 2.2 percent means serious problems — a health index if ever there was one.

When the potassium goes down with calcium going up, the equation changes immediately. In terms of short-duration grazing, potassium starts high and goes lower. Calcium starts low and goes high.

The two most important minerals in blood are potassium and sodium. That is why adding salt and calcium to such pastures delivers a cation to drive the potassium, with sodium and perhaps sulfur replacing some of that potassium uptake, with cows eating perhaps 15 percent more forage.

The equation now flowers into the use of vitamins, enzymes, even traces on the pasture. The texts all say that pasture staggers are due to magnesium, when in fact calcium seems to be the chief problem.

All the above leaves unanswered the bloat problem usually blamed on legumes. Yet when brix levels are up, animals will not bloat unless they are first starved, then overfed, because when the sugars are there, there is no environment for bloat to happen.

Book Three

The Futurists

21

Biodynamic Pastures

In the 1920s, the philosopher and clairvoyant Rudolf Steiner gave a series of lectures on agriculture in the Silesian estate of Count Keyserling. Invited were agronomists who were concerned about the faltering productivity of their land. Steiner's audience was stunned to hear him say that fresh cow manure from lactating animals packed into a bovine horn, then buried in the ground during the appropriate phase of the moon during winter months could transmogrify into a sweet-smelling dough-like substance teaming with millions of microorganisms. A single ounce of this substance diluted in water, then alternately stirred clockwise, then counter-clockwise for a given period of time would produce a spray with remarkable properties. Sprayed on dead and unfertile land, this material was capable of bringing it back to fruitfulness. Central to this process was Steiner's allegation that "etheric formative forces" in nature worked to metamorphose and consecrate cow dung. One of Steiner's followers, Gunther Wachsmuth, wrote a fascinating book called *The Etheric Formative Forces in Cosmos, Earth and Man.*

Then, as now, so-called conventional scientists scoffed at Steiner's recommendations as those of a man who had parted company with his senses.

Biodynamic Insights

• Plants not allowed seasonally optimum growth on top will gradually also shorten in roots, and sod-bound shallow roots result.

• Microbial and worm activity ceases below root levels.

• The need for any soil ripping is a result of bad farming. Mechanical ripping is costly and will only temporarily relieve the compacted conditions.

• Ripping wettish clay only causes cutting and smearing of soil pores where cut, thus inhibiting air penetration.

• The biological equivalent to European winter for Australian soils is the state of soil hardness and inactivity of extreme dryness.

• Soil, when it is truly biologically active (as in photo no. 2, structured soil), alone can account for produce with natural quality.

• When there is an immediate requirement for major or minor nutrients, judicious biodynamic procedures would get these major or minor elements out of fertilizer bags in the conventional way — relying on a sheet composting action through Preparation 500 in the soil — rather than running the risk of introducing concentrations of undesirable chemicals into the soil, via dubious so-called "organic" manures, city refuse, etc.

• The important thing about a humus colloid is that the soluble elements in it are at all times available to the plant, and yet they will neither evaporate nor leach out. Humus will hold 75 percent of its own volume in water.

• A plant actually grows from the leaves downward as well as upward from the roots.

• Artificial fertilizers are not poisons . . . but they actually become available in a form which is outside the organization of nature. That, of course, is rather dangerous.

• So-called organic manures in most cases are of highly dubious value, and these include chicken, feedlot and various stable manures. Some of the stuff coming out of the oceans near the shore is full of heavy metals and pesticide residues. These materials are invariably applied in a form that permits water soluble elements to remain water soluble, and do not constitute real compost in the sense of Preparation 500, where a total has developed. There is as much danger in water solubility problems from these organic manures as there is from factory acidulated salt fertilizers.

• Bloat does not occur at all on biodynamic farms. Neither does sterility, acetonemia, etc.

• We are farming within a vast cosmological and earthly environment. After all, our earth is also a cosmic body, and we must not forget that fact. We are a little bit too tied down to the fact that this is a material earth, and everything seems solid. We are inclined to think that something even as important as sunlight is a bit mystical, because it comes from far away. It is not at all. We are, in conjunction with other cosmic bodies, one huge cosmic ecology. If we work within that vast environment, then we are working within the organization of nature, and that would be true biological farming, not just using organic manures or green manures.

• At first when you sow down clover, you are not going to build up nitrogen, you are going to lose nitrogen in the soil. If peas and beans and vetches and many other legumes have been sown, and have been allowed to grow just beyond flowering stage, even if they have not formed a decent pod yet, they will be found to have used more nitrogen than they leave, if they are mown down or plowed in at that stage.

Although biodynamics has been around for decades, few U.S. farmers larger than extended gardeners stayed with the program long enough to understand the strange alchemy involved. An Australian film on biodynamics made the rounds under the auspices of Christopher Bird — co-author of *The Secret Life of Plants* — and with these showings, Alexander Podolinsky became an icon of discovery in pasture management.

Podolinsky is a Russian count who was brought up in Germany, then immigrated Down Under in 1947, almost immediately after World War II. He brought with him Steiner's teaching on a scale

Compacted soil.

so large it put to bed the idea that biodynamics belonged only to dilettantes and backyard gardeners.

The film set the parameters with a slant that those who saw it would not soon forget. There were images of cows at pasture in an early evening on a dairy farm in Victoria. Conditions were cold and damp. The phase of the moon was important, the commenta-

Structured soil.

tor said. Planet Earth was ready to receive a blessing, a spray that would make the pasture grow while at the same time transforming the soil.

Not many people can identify with the vision of a seer, yet one of the greatest discoveries ever did not flow out of experimentation, but issued forth when Albert Einstein simply exercised brain power to reason that $E=mc^2$, the rubrics of energy and mass and the speed of light eluding all but the most educated. Steiner's vision was of that order.

According to Steiner and Podolinsky, biodynamics taps a universal force so powerful, it supports all life. "In biodynamics," the film noted, "the universe must remain in a state of order and harmony."

That harmony is based on certain principles, all of which must be stated openly before a student can take seriously the system and its results.

Podolinsky gave me these premises at his farm, and they went quite a way toward explaining a system almost unknown in the United States, yet as common in Australia and New Zealand as popcorn in a movie theatre.

The above described balance is called the *cosmos.* Much of this is presented as spiritual in the documentary, which explains that the forces of nature are contained, albeit dependent on human understanding of living things.

"If I'm ever not in a good mood in the morning," Podolinsky said, "I very soon am again once I get near the cows. They tell me exactly what I'm like, and they notice cosmic happenings. A cow relates to the whole environment. To live in harmony with your environment requires a purity of spirit. Cows live in the total environment, not just in the pasture or on the farm. They can teach a scientist where else to look."

The disturbing fact is that scientists have no explanation for the successes of the biodynamic pasture or farm.

There are farms Down Under that have dairy cows for the purpose of filling cow horns with fresh manure. The creation of a product called BD500 is absolutely central to taking the world's worst soil and turning it into the best soil. Fresh manure loaded into cow horns with the deftness of a vendor filling an ice cream cone is only one episode in the year-long drama that puts quality grass on the ancient and tortured soils of Australia. The horns — thousands of them — have to be buried in a pit, and when the biodynamically alive product is harvested, it has to be stored correctly, and the pasture spray made therefrom has to be stirred and applied in the cool of the evening. Manure is no longer manure long before the stirred product is ready for the pasture, crop or soil.

This much stated, I would now like to recapture the essential elements of a discussion I had with Podolinsky. Some of his remarks are paraphrased, others are quoted directly, all are straight-from-the-shoulder insight, and all remain one of the best-kept secrets in producing grass for animals fed strictly on grass.

First, there are several "preps" in the biodynamic system. They will be described in detail in Chapter 22. For now it is enough to note that these preparations operate in a manner analogous to homeopathy, and are numbered from 500 to 508, typically referred to as "BD 500," etc.

Steiner and Podolinsky actually represent first-rate science, and those who view biodynamics as some sort of mysticism are merely codifying their ignorance. Few would characterize Einstein as a clairvoyant, instead he is seen as a scientist who could conceptualize a part of the Creator's plan. Similarly, when Rudolf Steiner conceptualized the formulation principles that led to biodynamic farming, he discovered realms of nature others would not comprehend for decades.

All the preparations are the products of reasoned science dovetailed together with experiments and practice. I can sketch a few principles, but in the final analysis there can be no substitute for reading Podolinsky's books, Lectures I and II. Without a clear understanding and obedience to nature's ruler, reliance on a prep is useless.

In his book *Biodynamic Agriculture: Introductory Lectures*, Podolinsky brought the problem of soil systems into worldwide focus with two startling photos, both of which are reproduced in this chapter. The question here is more real than rhetorical. How does the biodynamic system change the soil profile — soil with only a few thousand microorganisms per gram of soil — into the soil depicted in the latter of these photographs?

Harvey Lisle, the author of *The Enlivened Rock Powders*, joined me in viewing Australian soils that were much worse than the compacted example reproduced in the accompanying photo. In parts of South Australia the rainfall is much lower, and yet lush pastures are created with the help of BD500. Problems encountered cannot be dealt with via the agency of rippers, plows or conventional fertilization. The soils are sandy and low in organic matter. There is little more than fungi in the soil, and no microbes, certainly no worms.

Even with 8 or 10 inches of rainfall, only 2 inches sink in because the soil has become non-wettable. Such soils reject water. There are hardly any trees. When it rains, too often the moisture is sucked away by the winds. In most poor-rainfall areas, the moisture comes in heavy, short bursts, never as a small penetrating rain. When such soils are changed from dead to live, plants thrive immediately, roots penetrate more deeply, and moisture is absorbed in the new humus. Then plants get their 8 or 10 inches of moisture.

With a tape recorder running, Alexander Podolinsky connected the dots in the unique picture we saw: The session was hurried, yet in depth.

"Mr. Podolinsky, it appears you have a time problem uniquely your own. You run an active farming operation. You write. You consult across a nation with a land mass as big as the United States." In response, Podolinsky hurled off a lecture that explained many of the exquisite pastures exhibited in Australia and New Zealand, none of which could be produced in the United States using conventional or even legislated organic methods.

"Every word I speak for the rest of my life is to be turned into a deed," the farmer-scientist said. He said he speaks to practical men who want to change their farms. As for any other type, "I don't speak to him at all. Instruction is useless if this man is not willing to follow instructions."

The changes available can be seen firsthand on farms too numerous to catalog here.

Podolinskly was adamant. "A man who does not do everything right cannot be a biodynamic farmer. There are people who are interested in selling iron. If you use a rotary hoe or disc hoe, as you do in the American Midwest, you loosen the soil too fast. You make dust out of it — it all scrapes down to one level. The water falls through and stops, creating your hardpan. The chisel plow, on the other hand, pulls soil apart. The Graham-Hoeme plow — which came out of Amarillo, Texas — is the only American implement we have used here in Australia. People don't understand why they call it a *chisel* plow — they all treat it like a tine cultivator. In a cultivator or an agri-plow, the shank is one way. With a chisel plow it's the other way around. As a consequence, the chisel plow has springs, not just in the mechanical spring itself, but in the arm. Its shank can be tense, but the other type can't be tense — it'll either snap off or will stay exactly in the way, because it is fixed. This means the chisel plow — if it is tensed right, you have it deep enough, and the soil gives enough resistance — *hammers* underground. It does what the old Chinese market gardener did. They took the topsoil off, put it to one side, then they took a pick and loosened the clay underneath. Finally they put the topsoil back. They'd go around the paddock that way. Only the chisel plow can do this. As a consequence, you do not build a hardpan. The Wallace plow has a foot at the bottom. It cuts a level surface the same as a disc plow. The moldboard plow kept the soil going for a thousand years, with a long shank in the front, but pointing down so it pulls the soil off the bottom. The soil is broken open, it is not cut. It doesn't get really hard like a rock — all the pores are open, and water goes through. Air also goes through, and that is vital. The Wallace plow is flat. It has a complicated point. They can plow about a half a day with it, at which point they have to buy a new point."

This Australian master of pasture and crops does not charge his clients for going to their farms. He assiduously avoids making a profit from the preps he provides because, in his words, "If I charged as they do in New Zealand, I'd make between $1.5 and $2 million Australian. Steiner *gave* us Preparation 500. Some suggested it be patented. He said, 'I won't — I don't want profiteering from it.' If I paid a lot of tax at the end of the year, I would feel guilty. What money I make, I use. I give some away. I don't want to pay much by way of taxes. It will go into armaments or bureaucracy, neither of which I can support."

In sounding out the pasture story, the two plots reproduced above returned into focus as this great agriculturist continued. Some of the sugarcane country provides a good horrible example. There is no soil worse than soil in banana country or cane country. These soils have received the most fertilizers and the most poisons. Cane country is all bleak, flat soil, literally as hard as the top of his desk, Podolinsky said. If you slowly work a fork into it, you can't pull it out again; it has that much glue power. Yet Podolinsky has a video illustrating the transformation of that soil within one year. In some cases it takes a bit longer. The job cannot be done mechanically. A ripper is only an excuse for bad management early on. The ripper does not make soil. It does not make structure. Only roots can do that.

What, then is the approach? "I go from property to property. I would want to know your age," Podolinsky said. "I would want to see your finances and equipment. I always put myself into that man's boots. I rely on my own experience or I couldn't advise honestly. You have to make it all workable," whether it's crop, woodlot or pasture.

Obviously, the farmer who is in debt has to crop quickly in order to lift the burden. Otherwise, cropping should be 25 percent of the acreage, with 75 percent pastures supporting stock. If the soil is that of picture number 1, then ripping is essential. When you go through the hardpan with a crowbar, it is moist underneath. Often a 300-horsepower tractor is required.

First, the soil has to be dry so that it will bust up into big chunks. Few farmers correctly accomplish ripping. The procedure is something akin to pouring a cement slab and then topping it off with soil to grow a crop, any crop. In two years, that topsoil will be hard. The same is true for soil that is ripped for wheat production. Wheat has no root permanency. Under the Podolinsky biodynamic system, "If we rip, the land is not cropped immediately. For three years the farmer has pasture on it, whatever sort of pasture that part of the country will grow."

Pasture *must* have sheep and cattle on it. This way the soil is improved permanently with a root structure that is deeper and more profuse than is possible with wheat. Next, gentle plowing — never faster than a horse could walk — makes cropping possible. Quite often the biodynamic farmer no longer wants to crop, for he quickly learns that he can make better money out of his pasture. The pastures become so successful, many farmers have to be pushed back into needed grain production.

Grass is the key to soil number 2, and microorganisms are the key to grass.

Rice is a crop that can thrive in a biodynamic approach. It requires very good soil in order to germinate the seed — of all grains, rice is the poorest germinator. That's why the old Chinese farmers would seed boxes to accomplish germination. Then they would hand-sow seedlings into the water. Even for China, such a procedure is too labor intensive today, yet sowing into dead soil is equally frustrating. Thus, the paddies are filled with water and pre-germinated rice is dropped in from a plane. Unfortunately, rice production is bedeviled by a little red worm that eats the root so that the plant comes loose and floats on top. Growers use chemicals to deal with this pest. Those toxins enter the irrigation ditches, then the ground water.

The biodynamic farmer, on the other hand, relies on pasture and on sheep to graze it down — "then we sow the rice into that," Podolinsky explained. "We water it, and we spray Preparation 500. The rice germinates and grows up with the pasture. We put the sheep on the second time for grazing. By this time the rice is strongly established. The cereal grows up more quickly than the clover. When the rice is sufficiently high, we put on the water. The clover can't grow anymore." What about that little red worm? Only a few seem to thrive, and their destruction capability is minimal. Intervention with toxic chemicals is not indicated.

Thick pasture inhibits the growth of weeds. In Australia's fragile terrain, fertilizers become superfluous when grass dominates and row crops are produced about every seventh biodynamic year.

This concentration on returning carbon back into the soil and developing nitrogen release is an article of faith among biodynamic farmers Down Under. The crop is thus energized. With rice, once the water is off, the pasture comes back by itself. Even after several months of standing water, the pasture comes back.

Pasture without stock is close to being an oxymoron. Pasture thrives only in conjunction with stock. In a manner of speaking, grain growers are merely soil miners and ought to be given a depletion allowance on their income tax, except that the real farm is not a mine. To have a real farm, according to biodynamics, you have to have livestock.

Pastures ask for BD500, and BD500 means horns of manure buried according to the esoteric rules of the visionary. The pit has to be as big as a Cadillac car and horns have to be numbered in the thousands. If the internment is not crowded, roots will invade the underground sanctum. Those who want to get serious about pastures cannot take a report such as this one to be a working manual. Homework is required, and this means reading literature many without the will to understand denigrate as "underground," in the pejorative meaning of the word.

There is no shortcut. Nature built its Texas pastures over centuries. Biodynamics can do the job much faster, the Prep 500 first being produced by an expert, and the single ounce being spun into a complete spray mixture.

Professors shake their heads in abject disbelief at this business of setting up a vortex in a vessel, reversing the action to achieve chaos, then more of the same for a prescribed time period. Nothing in so-called settled science explains what Steiner saw, and possibly only a few masterminds understand all of it to this day.

Mere mulch will not cause roots to go down. For this requirement to be met, there has to be biological activity. Something has to come up and something has to grow down. As deep as possible.

In any case, mulching should never be done with alien materials. Left in place too long, it stagnates.

In biodynamic pasture management new roots are produced all the time. They evolve into humus when the plant grows up quickly and is eaten down before it joints, activity is stimulated. Once rich pasture is achieved, it has to be mowed or grazed at optimal length.

Rotational grazing management granulates the soil. In any case, the physical size has to be right if Prep 500 is to work right in plants. The physical side is fully half of biodynamics. *Right* has a real meaning, as Podolinsky illustrates. "I take on a wheat farm. The man says, *You will want me to keep the trash on the soil to plow it in.* That is what many people call organic. But the answer is *no!* The soil has no microbes, certainly no nitrogen. If you spray 500, you'll get some microorganisms, but for their maintenance on such poor soil they will use the least bit of nitrogen to break down that old straw. Otherwise it can't be broken down because they must have protein. Until that soil is correct, we don't plow the 'trash' in. In such a case it is better to get a quick burn and get some elements on top."

In the case of cane, it is harvested green, never burned.

There are soils so poor they can support little except lucerne alfalfa. If it lives more than a year, roots will go deep enough, perhaps 12 feet. There is nothing in such sand to feed much of anything without such a root system. Under such circumstances fertilizers are mandated, as are trace elements. Side dressings of rock elements are indicated. Once the biology becomes operative, the nutrients become available.

Rock powders and BD500 twice a year changes soil in two years, and fescue put depleted acres well on the road to recovery. "In six years we've been successful on whitesand farms to a point where two acres will carry a big Hereford cow plus calf until the calf is a yearling." Podolinsky said.

Withal, good preps are hard to come by. Ernest Jacobi in southwest Germany is a producer ratified by Podolinsky. Austria has preparers. There are others. The Josephine Porter Institute (Woolwine, Virginia, phone (276) 930-2463, e-mail <info@jpibiodynamics.org>) has a small clientele in the United States, the one nation that needs more pastures and less row crops, yet seems mesmerized by the Svengalis of rescue chemistry. An excellent book on biodynamics in the United States is Hugh Lovel's *A Biodynamic Farm*, as mentioned above.

Podolinsky summarizes, "The so-called Anthroposophical organizers of biodynamic agriculture you see in most countries are not practical farmers and scientists.

They often say everyone should make their preparations. That does not work. Usually they do not know how. It takes years to learn. I make more 500 in one year than the rest of the world together has made since 1924."

The spraying has to be done at no more than ten pounds pressure. Why does it work in Australia and New Zealand and not elsewhere?

It takes an experienced hand to supervise the initial foray into biodynamics. The storage of 500 has to be right — not refrigerated, not in plastic, not dried out. The failure rate in the Mecca of biodynamics, Australia and New Zealand, is practically nil.

The biodynamic road to top-notch pastures is governed by geography. This makes rules somewhat malleable. There are nuances, but there is always cow manure from lactating cows. It is turned into a product that when liquefied can be drunk by a cow. This humus substance is blessed by the cosmos after only three months. Steiner said this had to be done over the winter. In Europe the soil is frozen immediately in winter. If you dig nine or ten yards deep, it is summer down there, winter on top. In the Podolinsky pit, only two feet of soil insulated the horns from ambient air. The horns full to the brim lie atop each other, but not touching. Good soil between them provides insulation. Down Under, the winter temperature hovers around 42 F. Microbial activity is near zero. Green manure becomes brown humus in three months.

No scientific explanation attends the change during a non-biological time of the year. This horn material at 1.25 ounces per acre challenges all styles of eco-farming while leaving unexplained why only cow horns suffice. In any other vessel, 500 does not happen.

There is a summary of sorts to this look at biodynamics. Microbial activity, humus formation, and root activity are the same thing. "It is one activity viewed from three different angles."

In fragile soils it is often necessary to sow into existing pasture. The term here is sod-sowing. The idea is to have roots down deeper. Some of the natives won't go deep.

In Europe a rye grass can be treated with 500, putting microbes into the soil and giving them something to eat. This is not possible when the country is so dry the sun would take away the required moisture. In most of Australia the excess feed has to go into the stalk. Everything in a plant that is combustible, that can be burned with carbon dioxide in the air. Only roughly 3 percent is minerals. The rest comes from the leaves. All the world is constructed from carbon dioxide from that air. Now you plow it. If you do not have moist conditions so microbes can work, the sun takes it out as straw fire. Irrigation might serve that European intent, but it costs so much it makes grazing mandatory.

"On my farm," the Down Under biodynamic expert recalled, "We have a black sand with virtually no humus." The organic matter level in the top four inches initially was 0.9 percent. Where I used biodynamic preps after six years — no compost, no fertilizers — by rotational grazing the change was to 11.4 percent at various levels. Now at 40 inches, there was 2.4 percent, five times what was originally on top. The chromatograph of biological activity at 40 inches was higher than at the top 4 inches. This works especially in areas where top-down green manuring will fail.

Cows, horns, pastures, in that order and in that order of importance captivate the imagination. Some 130,000 horns, each filled by hand and emptied by hand, represents a lot of acres at 1.5 ounces per acre.

Manure is not manure, a tumble bug might say. BD500 was created with the knowledge that the cow is a unique animal. She has several stomachs, a refined digestive system. She chews her cud. She rarely runs. She is as calm as conditions permit. Her manure is 25 percent microbes. It enriches and enlivens the soil.

Outside the cow horn the manure does not convert. We do not know why.

Cows can be and are dehorned, but afterward they are never the same. They become dull and listless. The horn is a vessel of nature, the Australian film explains, and it has to be put underground to tap into water forces. It's the earth pull to which the horns are subjected.

The reader who wants to find grass and only grass in these lines may become frustrated, even impatient when insight at variance with the accepted norm is presented.

To this the answer is simply, All life is grass! And when it comes to grass production, Australian and New Zealand acres have yet to be equaled, especially those acres called biodynamic.

22

Lessons from Steiner & Pfeiffer

In closing this survey of the pasture enigma, it becomes evident that the seekers and the grazers are one. One of my best interviews over the past 35 years fielded answers that remain some of America's best-kept secrets. They enlarge on the wisdom of Alex Podolinsky and install an American slant into a discipline that cries out for recognition from a planet badly in need of repairs.

Hugh Courtney is a native of Chicago, a much-traveled son of a professional Army Corps of Engineers and Air Force father, and finally a student of the life forces that are a part of biodynamic agriculture. First he examined traditional organics, and then he became a seeker of the enigma — the goal a bit uncertain. At that time he stumbled onto Rudolf Steiner's *Agriculture* lectures, a book he now sees as a bit too difficult for the novice. Nevertheless, the door flew open when he caught the comment Steiner and Ehrenfried Pfeiffer made about the need to use BD Preparations widely in the earth for the healing of the Earth. It was a case *of Eureka, I've found it!* He made the idea *heal the Earth* his own. Putting the soil in a better condition was not enough. With a background of esoteric disciplines in tow, Courtney absorbed Steiner's terminology. Courtney has now retired, but BD Preparations and information concerning the subject matter of this chapter can be obtained from the Josephine Porter Institute.

Q. What was Steiner's background?

COURTNEY. Steiner was a philosopher and scientist who had a fairly distinguished reputation in science. He was asked at a very young age, after he graduated technical school in Vienna, to serve as the editor for the scientific works of Johann Wolfgang von Goethe, who was preeminently known as a playwright, the author of *Faust*. He was also a poet and man of letters rather than science. But Goethe himself felt that his greatest contribution to the world was his scientific work. He is responsible for a theory of color that Steiner spoke of on several occasions — it is counter to the Newtonian theory of color, in which light is made up of colors, with white light being a collection of other colors. You find this out using a prism. Goethe approached it 180 degrees out from that and said colors are a breakdown of light rather than light being composed of the various colors of the rainbow.

Q. Was this knowledge of agriculture that Steiner had intuited?

COURTNEY. Steiner was a seer, that is, someone who could see other dimensions, if you will. He was born with a certain clairvoyance, but he schooled it. He subjected it to constant verification because a great danger of clairvoyance is that you will find yourself wandering in illusionary spaces. Steiner schooled himself very carefully so that he eliminated as much as possible the illusions that can come about, and, in essence, he arrived at some of his answers by holding conversations, if I may phrase it that way, with spiritual beings. When the farmers, veterinarians and so forth came to Steiner with the problems that they were experiencing as a result of the move toward artificial fertilization that took place at the beginning of the century and beyond, one can imagine that Steiner held a conversation with the beings of nature that could speak to him about what was happening as a result of man's agricultural methods.

Q. He knew how to interrogate nature. Is that what you are saying, in effect?

COURTNEY. Yes.

Q. And he was able to relate to these farmers an approach that would get them past the push toward artificial fertilizers?

COURTNEY. Before he actually gave the lectures, he was urged many times to present them, and it is as though he needed to prepare himself first by a deep study of what was necessary.

What he came up with was given in the lectures, and he presents a number of incredible ideas. Steiner saw the farm as a living organism. The farm individuality is an idea of his that is frequently referred to. Beyond that, he sought the answers for the problems that were presented, and the answers that he came up with were what I refer to as the nine biodynamic agricultural preparations. These are BD 500, 501 and so forth.

Q. What exactly is preparation 500, for example?

COURTNEY. In very simplistic terms, BD500 is cow manure placed into a cow horn and buried during the winter months. You take what can sometimes be a very smelly, even an offensively smelly, substance — excrement from an animal — and you pack it into a cow horn and bury it, and it comes out sweet smelling, full of micro-life and quite transformed in color, usually a very dark brown or blackish color. When one empties it from the cow horn, if you are sensitive at all, there is a certain radiative force that emanates from this substance. We take approximately one-quarter of a cup of that material. This amount is what we call a "unit," and a unit will treat an acre. We stir it in three to five gallons of water and spray it out over an acre. Over a period of time, not necessarily with just one application, it will transform that one acre into very well-structured, humus-rich soil. It is difficult for us in this day and age, with our education focusing so much on the material, to envision that such a little bit of substance applied in such a way could create this transformation and even create a very substantial deepening of topsoil over a period of years.

Q. In order to make this solution, you have to stir it in a certain way, do you not?

COURTNEY. Yes. It is placed in the water and stirred with a stick, by hand, or occasionally machines are used. You create a deep vortex and then reverse direction — you stir counterclockwise and then clockwise, etc. You create a deep vortex in one direction, then you reverse and create the chaos until you can stir in the opposite direction and create another vortex. You keep reversing directions, and you keep it up for one full hour. That's the stirring process that you use for the 500 and the 501. The 501 is ground quartz or feldspar placed into a cow horn and buried during the summer months. A very small quantity of it, half a teaspoon or less, is placed in three gallons or more of water and sprayed out as a very fine mist over the same acreage. The 501, however, is more appropriately viewed as a fertilizer for the atmosphere. You create this fog-like fine mist in the atmosphere so that the plants are immersed in it, so that they can breathe it in. The principle behind the stirring is that there are certain forces within these two preparations that are transferred in the stirring process to the water molecules. When it is sprayed out over soil or atmosphere, each droplet contains those forces encapsulated. The plants and the soil can, in essence, breathe in or digest — whatever you want to call it — those forces.

Q. Do you spray these preparations in the morning or in the evening?

COURTNEY. Generally the BD500 is viewed as an evening spray, applied near or even after sunset. The BD501 quartz silica or felspar is sprayed in the morning, at sunrise. You have to get up early and get it stirred and applied — hopefully right at sunrise. Two things dampen people's enthusiasm for biodynamics right away: one is this hour-long stirring, and the other is getting up early enough to stir and spray by sunrise, especially in the warmer time of the year.

Q. But it does nourish to the point that the life in the soil really gets to perking, is that right?

COURTNEY. You create beautiful soil structure, and the micro-life in the soil really blossoms. You can find earthworms where none were before.

Q. I saw a section of soil in Tasmania where the farmers had built their clubhouse out of adobe. That same soil was turned into a garden with biodynamic preparations, and you could run your arm in up to the elbow.

COURTNEY. If you are diligent in your application of the sprays and the making of compost with the other six compost preparations, then you will achieve beautiful soil.

Q. Let's talk about these other six compost preparations.

COURTNEY. We have 502 through 507, which are used in making compost. BD502 is made from yarrow blossoms placed in a stag bladder and hung in the summer sun. When fall comes, it is placed in the earth. A year after first being placed in the stag bladder it is exhumed, and the finished product is what we call 502. You can still see the structure of the yarrow blossoms, but they have transformed considerably. BD502 is especially useful for stimulating the potassium forces. In biodynamics we talk not so much about substances as we do about forces. We need to understand the effectiveness of biodynamics as something that is arrived at by stimulating forces. We are not putting 10-10-10 on, we are stimulating forces within the earth and atmosphere.

Q. Forces that are locked up for one reason or another?

COURTNEY. They are dormant, locked up, however you want to express it. The next preparation is BD503, the chamomile preparation. Chamomile blossoms are placed into a bovine intestine, that is the gut of the bovine species, and that, again, is buried in the earth during the winter months and then exhumed. When it is dug up and processed, there is almost no flower structure left to be discerned. Chamomile is particularly effective for calcium processes. The next preparation is stinging nettle preparation, called BD504. This one requires no animal parts, you simply bury it in the earth. The nettle is usually harvested in June here, and we bury it so that it spends the winter and the following summer in the earth — spending 15 months or so in the ground. When it is dug up it is usually a very dark black.

Q. What kind of moisture level are you looking for?

COURTNEY. With the stinging nettle, you don't add any extra water. With the yarrow and chamomile preparations, you moisten the blossoms sufficiently to make them easy to insert into the bladder or the intestine — just lightly moistened, so that it has a certain slipperiness when it goes into the sheath. The stinging nettle is a plant that is very rich in iron, and it has a very health-giving quality to it. It makes the plants sensitive to their surroundings and has a lot to do with nitrogen effects, as well.

BD505 is the oak bark preparation. In this country we use the white oak; in Europe it is generally the English oak. If you are in certain parts of the country where the white oak doesn't grow, other oaks could be used. You take the very outer layer of bark, grind it up, and insert it into the skull cavity of a domestic animal. You bury it in a swampy place or in a rain barrel in sawdust or plant material so that it spends the winter in that environment. Then you dig it up and remove the oak bark, which is particularly effective against plant diseases. Our experience requires us to use a fresh skull, and that is a bit off-putting. We have found that people don't want to deal with the gore necessitated by some of these methods.

Q. But they don't need to, they can get the preparations from Josephine Porter Institute, can't they?

COURTNEY. They can, but ultimately people need to approach the making of the preparations themselves to the maximum extent.

Q. Is there one more preparation?

COURTNEY. There are two more. We have the dandelion preparation — dandelion blossoms placed in a bovine mesentery or peritoneum/mesentery tissue. Peritoneum is the tissue that holds all the organs within the abdominal cavity. The mesentery, or the isles of mesentery, is the tissue that holds the intestines together. It is basically just one continuous tissue that folds in and back and winds around. It is an architectural marvel. The dandelion preparation brings certain calcium silica forces to the compost.

The last of the compost preparations is valerian, which comes from the juice of the valerian flower. It is not buried in the earth — it is made into a wine, if you will. It goes through a fermentation process, and a very few drops are then placed in water and stirred. It is then either placed in or sprinkled on the compost pile, depending on what school you come from. It brings the phosphorus process into the picture. These six biodynamic preparations — yarrow, chamomile, stinging nettle, dandelion, oak bark and valerian — are used in making compost.

Q. They are inserted into the compost pile as individual units?

COURTNEY. Yes. A single unit of most of these is approximately a level teaspoon in quantity. That teaspoon will direct the breakdown of the material within the compost pile and then the build-up into a new structure.

Q. How big a compost pile are you talking about?

COURTNEY. Upwards of 10-15 tons.

Q. So you don't need much of it to migrate through the entire pile?

COURTNEY. Steiner used the word "radiate." The energy radiates through the entire compost pile and directs the substances in the breakdown and buildup process. Instead of having a pile of smelly, ammonia-gassing material lying there, you end up with more nitrates in the actual pile.

Q. What sort of a pile are you looking for in terms of carbon and nitrogen?

COURTNEY. Good composting principles, which is a 15:20 ratio of carbon to nitrogen. One should still aim for that. With straight animal manure, for instance, cow manure, you don't need to add anything — you can compost pure manure. Steiner made a distinction in his lectures between a compost of plant material and manure compost. He used two different terms: "compost" and "manure compost." Manure compost was what one would have in a livestock or dairy operation where the manure is collected as almost pure manure with only a certain amount of bedding in it.

Q. Of course, if you got horse manure with a lot of sawdust in it, you would have a different composition.

COURTNEY. Yes, you would have a very high carbon ratio for most of what you can get nowadays for horse manure. In those instances you might actually want to do a few things to bring in a greater nitrogen effect, because you don't get that much from the horse manure due to the way they bed the horses these days.

By the way, speaking of horses reminds me that there is a third biodynamic spray preparation I should mention: horsetail herb *(equisetum arvense)*, sometime referred to as BD508. Many practitioners object to labeling it as a bona fide biodynamic preparation, but here at JPI, our experience teaches us that it truly deserves an equal ranking with any of the other preparations. While the orthodox view of horsetail herb sees it as useful merely in the prevention or cure of fungal diseases, we have come to realize that it is a very powerful regulator of the watery element, extremely useful in conditions of either drought or excessive rainfall. If there were widespread use of BD508 — in concert with the other biodynamic preparations, of course — I would venture to say that you would not have the devastating fires now being experienced in this country. The elemental beings would be able to conduct their work in a far more harmonious way. To cover the subject of horsetail herb fully would almost require another full interview.

Q. It seems as though one would have to have a great deal of emotional involvement with the love of plants and the soil in order to accept and adopt the esoteric things that you are telling me. Would that be a correct assessment?

COURTNEY. In part, I suppose I could agree with that. Steiner spoke of the three-fold human being: a being that brings thinking and feeling — the emotions — and will to his actions, his behavior, his living life, if you will. One can exercise the intellect, the thinking ability, a great deal in biodynamics. Maybe that is part of the problem with biodynamics, we spend a lot of time thinking about biodynamics instead of doing it. I identify four kinds of biodynamic practitioners. The first is armchair type, who sits there and reads the book but never does anything. Then there is the straw-fire type, who is on fire to do this wonderful, marvelous thing and is full of emotion and feeling about it and then he discovers that there is actual work involved — what! stir for *an hour?* Then you have the one who brings the will forces into play, and this is generally the orthodox type. He could say "this is the way my father did it, and this is the way I will do it," but he *does* do it, so it gets done. Generally, however, this type doesn't bring the heart and mind into the process. The ideal practitioner is one who tries to combine his three-fold being in the entire process of biodynamics. I suppose I could say that there aren't too many of us who manage to arrive at that state, and I should probably include myself as one who hasn't got there as much as I should.

Q. What is it that is making it so difficult for Americans to make this connection?

COURTNEY. I can answer that the same way that Steiner answered Ehrenfried Pfeiffer, who had much to do with bringing biodynamics to America. Pfeiffer asked Steiner why there was so little spiritual progress being made. Steiner's answer was that it was a matter of nutrition. Beyond that, he spoke of geographic forces in the earth. He spoke of America as a place where the hardening or materialistic forces are very much stronger. In Asia it is the exact opposite — the airy-fairy forces, the illusionary forces, are very much stronger. But here in America we have a great deal of materialistic forces to overcome. We are hardened in that respect. Modern education also hardens us in the way that we are taught. We don't easily access the spiritual world, and we don't grow easily to spiritual awareness in America. We definitely do not eat food that is endowed, if you will, with spiritual forces. This endowment comes about when the preparations are used. You bring in cosmic forces from the periphery, and you awaken vital earthly forces so that life exists within the plant. It doesn't just become so much substance that we stuff into our mouths and gut. If food that is endowed with cosmic spiritual forces is eaten, then we will awaken more readily to thinking along spiritual lines instead of along these materialistic lines, evidence of which abounds in such things as genetic engineering. We think we can mechanically manipulate the genes in the living structure of the world and arrive at something that will have lasting value. I think we deceive ourselves, and sooner or later that will show. At any rate, the answer is that we need to eat food that is strong with these forces.

Q. How did you get involved with biodynamics?

COURTNEY. Basically, I wanted to pay more attention to my health and started to grow things organically in a suburban garden situation. It didn't take me too long to realize that organics didn't quite do it for me.

Q. Did you want something more?

COURTNEY. Organics just didn't provide what I felt needed to come about in plants. About that time, I stumbled onto a copy of Steiner's agricultural lectures at a health food store in College Park, Maryland, where I was living. I opened the book and caught the comment that Steiner and Pfeiffer made about the need for these preparations to be used widely in the earth for the healing of the Earth. That was what I was looking for. For me, organics did not heal the Earth — it put the soil in a better condition, yes, but there was no healing taking place.

Q. It was not taking it far enough, in your opinion?

COURTNEY. That's right. I devoured that book of lectures. Everyone tells you in biodynamics that you never start first with the lectures, because you will be in such deep water that you will just put it down and not go back to it. Somehow that didn't happen to me. I had a bit of a background already in certain esoteric disciplines, so I understood a bit of the terminology. That was the first book of Steiner's I ever read. Some in the movement suggest that it is the most difficult of Steiner's works to understand.

Q. If you wouldn't recommend that book for students interested in biodynamics to start with, what do you recommend?

COURTNEY. I would really recommend going back and reading Pfeiffer, actually. *Soil Fertility, Renewal and Preservation* is still a classic. It still has one of the best chapters on farm manure composting that you can find anywhere. The book *Grasp the Nettle*, by Peter Proctor, is a perspective from New Zealand, but it is still very helpful. Podolinksy's books are also good, as well as Wolf D. Storl's *Culture and Horticulture: A Philosophy of Gardening*. Hugh

Lovel's book, the part about biodynamics, is good, too. A lot of the early books are good.

Q. Would you tell me a little about Ehrenfried Pfeiffer?

COURTNEY. Pfeiffer was a close student, or disciple, of Steiner. He came to America and helped get biodynamics started in this country. He devised his biodynamic products (the Pfeiffer BD Compost Starter and BD Field Spray Concentrate) in the '40s and '50s, when he saw that the American mind-set was not quite ready to deal with things like buried cow horns and stag bladders, cow skulls and so forth. He came up with his preparations as a more palatable way for people to approach biodynamics, and they are still excellent products.

Q. You catalog these products and advise people?

COURTNEY. That's right.

Q. Are they still making anything up there in Spring Valley, New York, under Pfeiffer's methods?

COURTNEY. All of that has moved down here, and we are the only ones making it now. Spring Valley is no longer operating.

Q. What if someone wants to get started in biodynamics, but doesn't want to make the preparations?

COURTNEY. If you want to get started in biodynamics and you are not prepared to make your own preparations, we will supply the preparations at the Josephine Porter Institute. Just contact us, and we will be happy to talk to you long enough to get you started.

Q. You were speaking of fogging some of the preparations. How do you get them to fog properly?

COURTNEY. We used to use a backpack sprayer with nozzles that can adjust from droplet size to a pretty fine spray. You can get sprayers from various places, and we do have some information on that at the institute.

Q. How much of these preparations do you need?

COURTNEY. One set of compost preparations, BD502-507, will treat upwards of 10 to 15 tons of material, whereas one unit of a compost starter basically does 1 to 1.5 tons of material. Both of those routes are available. Some people prefer one, some prefer the other. Both methods will yield very good compost. We have a fellow in Pennsylvania who makes quite a bit of compost, and his experience last year was that it was the best product he had ever found to make compost — he was using the starter we make here at the Josephine Porter Institute. He is an adherent of the Lubke techniques, and he has tried many other products on the market, but he felt ours were the best.

Q. Why is this so difficult for the American farmer? It is an interesting topic — you would think there would be more interest.

COURTNEY. We talked a little about this earlier. I think that people are fearful of the spiritual side of things. The biodynamic preparations in particular, in my view, present a significant (for lack of a better word) spiritual barrier to overcome. As I said before, by our geographic nature we have a great deal to overcome in our materialistic outlook on life. When explaining biodynamics, there is no other way to get around it. Sooner or later you have to refer to the fact that it has a spiritual dimension. Practical farmers and scientists don't want to hear that. Our foods today do not carry the forces necessary for us to awaken the way that we need to spiritually. Present-day agriculture, for the most part, produces food that is geared to keep us in our materialistic frame of mind. We have dead food. We eat substance without life force within it. Biodynamics is about reawakening the life force in the earth and in plants through the preparations. Chemical, even organic, fertilizing doesn't do that.

Q. Our current rescue chemistry certainly isn't doing it.

COURTNEY. Chemical fertilizers are not doing it, but I can be a bit critical about organic, too. I've compared the chemical approach in agriculture to allopathic medicine — if you are not careful, an overdose can kill you. It doesn't really cure anything, it just suppresses symptoms until eventually the organism recovers. It often recovers on its own, but seemingly the allopathic medicine has worked. The medicine only treats symptoms — it does not get to the root cause of things. The organic approach comes at things in a kinder way, without using harmful materials. But the underlying thinking hasn't changed, you are still thinking NPK. You want to bring a substance to produce an effect. You are still basically approaching things from a symptom point of view. I compare the organic approach to hospice care. In general, you are being very kind to the Earth, but you are not really curing it. A biodynamic approach is closer to homeopathic medicine. You are trying to effect a cure so that healing can take place. One of the first things I ever read in the agricultural lectures was the phrase "so the Earth can be healed.“ That is pretty much the motto of the Josephine Porter Institute.

Q. When you say “spiritual,” what is it that you mean?

COURTNEY. We talk about non-physical forces that emanate from the cosmos and from other beings of the Earth. Steiner talks about a whole hierarchy of spiritual beings that we are no longer aware of in our present age, but humanity did have this awareness for eons. Steiner was one of the people who came into the world with this awareness, but he schooled himself scientifically so that he would not be trapped by illusion. As I said earlier, one of the problems with clairvoyance is that one can convince themselves that what they see is the reality, whereas they are actually projecting their own inward illusions. Steiner schooled his clairvoyance along with his excellent scientific training. He was very conversant with what was going on in science in his time, including the work of Einstein, Freud and others. Steiner was also the editor of the scientific work of Goethe. Goethe not only came up with a theory of color, but also he developed theories of evolution that are very opposed to the Darwinian theories. Goethe is a fascinating individual, and Steiner understood him.

Q. You made a remark that our food is inadequate. Recently a farmer in Ohio told us that there wasn't a thing in the grocery store that was fit to eat. Would you comment on that?

COURTNEY. I am not sure which grocery store he went to, but I would probably endorse that comment. There is very little in the stores that is fit to eat, in all honesty. I no longer eat beef or chicken from the grocery store. The only beef I eat is what we raise here. The only chicken I eat is what I get from Joel Salatin. That is my own preference. I try to get as many organic and biodynamic vegetables as I possibly can. We do some modest gardening here. This is a farm, but I've never been so presumptuous as to call myself a farmer. I think you kind of have to be born into it.

Q. But you do make all the preparations right there?

COURTNEY. Yes. That is the main purpose of the farm. The preparations are the main product. Beyond making them here, we do whatever we can to teach people to make their own. We hold workshops here twice a year, and also I go to conferences and meetings and try to give demonstrations and workshops on preparations.

Q. You are working to get the information out there in the hope that people will just give it a try, knowing if they try biodynamics they will be impressed with the results?

COURTNEY. Yes, that's our hope.

Index

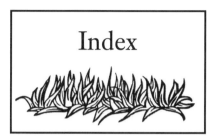

Also from Acres U.S.A.

Reproduction & Animal Health

BY CHARLES WALTERS & GEARLD FRY

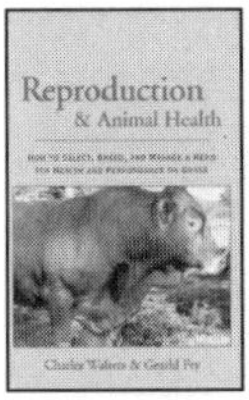

This book represents the combined experience and wisdom of two leaders in sustainable cattle production. Gearld Fry offers a lifetime of practical experience seasoned by study and observation. Charles Walters draws on his own observations as well as interviews with thousands of eco-farmers and consultants over the past four decades. The result is an insightful book that is practical in the extreme, yet eminently readable. In this book you will learn: how to "read" an animal, what linear measurement is, why linear measurement selects ideal breeding stock, the nuances of bull fertility, the strengths of classic cattle breeds, the role of pastures, the mineral diet's role in health. *Softcover, 222 pages. ISBN 0-911311-76-9*

Hands-On Agronomy

BY NEAL KINSEY & CHARLES WALTERS

The soil is more than just a substrate that anchors crops in place. An ecologically balanced soil system is essential for maintaining healthy crops. This is a comprehensive manual on soil management. The "whats and whys" of micronutrients, earthworms, soil drainage, tilth, soil structure and organic matter are explained in detail. Kinsey shows us how working with the soil produces healthier crops with a higher yield. True hands-on advice that consultants charge thousands for every day. Revised, third edition. *Softcover, 352 pages. ISBN 0-911311-59-9*

Hands-On Agronomy Video Workshop

Video Workshop

BY NEAL KINSEY

Neal Kinsey teaches a sophisticated, easy-to-live-with system of fertility management that focuses on balance, not merely quantity of fertility elements. It works in a variety of soils and crops, both conventional and organic. In sharp contrast to the current methods only using N-P-K and pH and viewing soil only as a physical support media for plants, the basis of all his teachings are to feed the soil, and let the soil feed the plant. The Albrecht system of soils is covered, along with how to properly test your soil and interpret the results. *VHS & PAL format available, 80 minutes.*

To order call 1-800-355-5313 or order online at www.acresusa.com